Avoiding Apocalypse

How Science and Scientists Ended the Cold War

Avoiding Apocalypse

How Science and Scientists Ended the Cold War

Jeff Colvin

Winchester, UK
Washington, USA

First published by Chronos Books, 2023
Chronos Books is an imprint of John Hunt Publishing Ltd., No. 3 East St., Alresford, Hampshire SO24 9EE, UK
office@jhpbooks.com
www.johnhuntpublishing.com
www.chronosbooks.com

For distributor details and how to order please visit the 'Ordering' section on our website.

ISBN: 978 1 80341 198 9
978 1 80341 199 6 (ebook)
Library of Congress Control Number: 2022938087

A CIP catalogue record for this book is available from the British Library.

Design: Stuart Davies

UK: Printed and bound by CPI Group (UK) Ltd, Croydon, CR0 4YY
Printed in North America by CPI GPS partners

Contents

Chapter 1

The Mandate of Heaven

Soviet Communist Party General Secretary Mikhail Gorbachev was not seeking specifically to avoid the apocalypse of global nuclear war, always a possible endpoint of the Cold War conflict between the U.S. and the U.S.S.R., when he released Soviet scientist Andrei Sakharov from internal exile in December 1986. Rather, he was simply acting to end the worldwide scientists' boycott of the Soviet Union.

As soon as Gorbachev had come to power in the Soviet Union a year and a half earlier, he had set himself the task of restructuring and modernizing the Soviet economy. Corruption and inefficiencies in the Communist Party bureaucracy had hampered development of the Soviet Union's centrally planned economy. In addition, the scientists' boycott had effectively isolated the Soviet Union from technological innovations that were driving economic expansion and modernization in the democratic Western countries. Gorbachev recognized that reform and restructuring were essential to economic modernization. He further recognized that the full integration of Soviet science into the worldwide community of science, and the benefits this would bring, were essential to the success of his reform efforts. It was then only a matter of time for him to draw the logical conclusion that in order to reap the full benefits of modern science and technology he would have to end the scientists' boycott. To end the scientists' boycott he would have to take the bold step of putting aside all the old Communist Party arguments, which he himself had supported, of why and how Andrei Sakharov was a threat to the state and to socialism, and end the forced exile of the most prominent Soviet scientist. It is likely, though, that Gorbachev had no idea he was setting into motion a sequence of events that, in astonishingly rapid succession, not only ended the scientists' boycott, but led directly to the end of the Cold War and even to the end of the Soviet Union itself.

As Cold War historian John Lewis Gaddis has pointed out[1], "(a)uthoritarian states that attempt reform risk revolution: it is harder than in a constitutional system to find footing in between." Gaddis himself acknowledges that this insight does not originate with him, but goes back at least to Tocqueville's analysis of the origins of the French revolution. In any case, there were evidently limits to Gorbachev's understanding of this basic insight.

There were also limits to understanding in the West. Very few people in the West outside the scientific community understood then the linkages between the scientists' boycott and the ending of the Cold War, particularly how it was not only possible but likely to get to an ending that avoided apocalypse. Even now, approximately three decades (as of this writing) after the dissolution of the Soviet Union, fundamental questions remain unanswered. What is the linkage between the scientific worldview and the idea of democratic government? How, when, and where did this linkage arise historically? How did this idea spread to the Soviet Union? How was this linkage further expanded to include linkages to fundamental human rights and world peace? How were these ideas incorporated into foreign policy decisions taken by the U.S. and the U.S.S.R. during the Cold War? What role did scientists play in influencing these decisions on both sides? What constituencies in the U.S. opposed linkages, and why? What role did the nuclear arms race, and attempts to control it, play in the thinking on both sides on these issues? What prompted the scientists' boycott? Was it effective? How did it contribute to bringing a peaceful end to the Cold War? What role did scientists play in the initial military disengagement just after the dissolution of the Soviet

Union? Finally, perhaps the most important question of all as we go forward into a world facing even more apocalyptic threats to human existence: why did this disengagement fail to last in the aftermath of the Cold War?

To make sense of how the ideas, the worldview, and the actions of scientists could have had such a disproportionate influence and effect on the conduct and conclusion of the Cold War, it is helpful to go back in time before the birth of modern science to see how fundamentally different the world was then. Since modern science was born with the work and insights of Galileo Galilei in late sixteenth-century Tuscany, I accordingly began a voyage of historical exploration by traveling to the center of power of what was the largest empire on earth in the sixteenth century.

It took four days to walk there. Since there are no roads that go there, and if one foregoes the train through the Urubamba Valley that passes by the base of the mountain plateau on which the city is perched, the only way to get there is by walking. A difficult walk it is, too. Most of the trail is above 11,000 feet elevation, over three Andean mountain passes, the highest at about 13000 feet. Unlike mountain trails in the North American Sierras and Rockies, there are no switch backs; the trail follows the contour of the mountain in a nearly straight path above the Urubamba Valley, the Sacred Valley of the Inca Empire; indeed, in some places, it was a steep climb on all fours.

It was, therefore, not difficult for me to understand, emerging from the mist as we descended the last mile or so of trail above the plateau, why the lost city of the Incas, Macchu Picchu, remained undiscovered by European and American explorers from the time of the Spanish conquest in 1532 until 1911. The place is invisible from the river valley below, and approachable only from the south, along the difficult route we came. On the other three sides is a very steep drop into the Urubamba gorge, and beyond lies the vast expanse of the Amazon rain forest. The American explorer Hiram Bingham stumbled upon the long-abandoned city in 1911 while searching for the capital city of Vilcabamba, the last refuge of the Incas. His discovery simply deepened the many mysteries about the last days of the Inca Empire, because it was, and still is, unclear whether or not Macchu Picchu is Vilcabamba, who lived there, over what period of time, or even what the city's main function was: a military outpost; capital of Vilcabamba; an imperial retreat; a health retreat; a training center for religious ceremonies; or a city of another people that was taken over by Incas retreating from the Spanish invaders. Important as these questions are to a full understanding of the Incas, the most important question for me was this: why did the Inca Empire fall so readily to the Spanish conquest despite the Inca's vast numerical superiority, and given their social, spiritual, and engineering development that was in many respects comparable to, or even superior to, that of sixteenth-century Spain? The answer to this question, I will show, bears directly on explaining the outcome of any clash of cultures, and in particular, how the Cold War ended.

When Francisco Pizarro marched into Peru in the Fall of 1532 at the head of a column of 62 cavalrymen, 105 infantry, and one priest, he had no idea that he was entering the world's largest empire. The domain ruled over by Atahualpa Inca[2] stretched along the west side of the South American continent from what is now southern Columbia to central Chile, and was home to some ten million people.[3] In contrast, the Kingdom of Spain, ruled over by Charles I, who was also the Holy Roman Emperor Charles V, and who had granted Pizarro the title of Governor and Captain-General of Peru, was only about one-quarter as big, even including modern-day Belgium, Holland, and Luxembourg, which were then part of the Kingdom of Spain. Even though the Incas had no wheeled vehicles, the Empire was linked by a vast network of roads, at least as extensive as those that the Romans built in Europe, over which the commerce and the armies of the Empire moved.

A strong and highly developed political, military, and social structure held the Empire together. Everyone spoke a common language, Quechua (conquered tribes were required to learn the language), and a well-organized system of taxes and tribute, in goods and services, supported the central government, with the Inca (ruler) at its head. In turn, the common people received the protection of the Inca. The Inca provided protection to the people by controlling the ritual structure around which their lives were organized. The objective of these rituals was to guarantee a good food supply, cure illnesses, and foretell events. The timing and duration of rituals were determined by a ceremonial calendar that was set by the Inca himself on the basis of astronomical observations of the positions of the rising and setting sun. The Inca, who was considered to be a direct descendant of the Sun God, and thus a god himself, carried out these observations from the central plaza in the capital city of Cuzco, located on a plateau at 11,000 feet in the Andes of central Peru, with the aid of carefully placed horizon markers on the surrounding cliffs.[4] Cuzco was thus the central engine of the ritual life of the Empire, with every person in the Empire driven by this central engine through the vast network of holy sites, or *huacas*, that were situated along radial lines from the center, and placed in the custody of individuals selected for their duties by the Inca. Thus, military occupation, for the most part, was not required to maintain central control, because the people believed that their very lives and livelihoods depended on their own support of the center through the *huacas*.

If Pizarro and his followers were unaware of all of this at the time they marched on Cajamarca, they certainly were not unaware of the high degree of technical and artistic skill of the Indians. The evidence was all around them. There were large public buildings and temples, finely crafted from large stone blocks that were carved and fit together without mortar so tightly that a knife blade cannot be inserted in between them. The size and weight of the stones, and the distances over which they had to be moved, suggests that the Indians made use of the same engineering tools the Europeans were using — rollers, inclined planes, levers, ropes, pulleys. The Spaniards saw exquisite metal work in gold, silver, and copper adorning the public buildings. They surely saw architectural sites that rivaled in magnificence the cathedrals they had seen in Europe.

Pottery and other goods were plentiful and factory produced with standard designs. The economy, as in Europe, was based on agriculture, with an organized food storage and distribution system in place that guaranteed freedom from starvation for everyone. Irrigation canals were built in the coastal plane to support farming there, and terraces buttressed with stone walls were built in the highlands to support farming there. Although the Indians never developed writing, they did have a number system, and kept track of their accounts with *quipus*, a set of knotted strings that is conceptually similar to the Chinese abacus and that can be used to add numbers into the thousands.

So, rather than encountering small bands of primitive "savages" in the wilderness, the Spaniards stumbled upon a large, well-organized, relatively wealthy, centrally controlled civilization that was at least numerically superior to their own. How, then, could they have succeeded in conquering and then destroying this impressive civilization?

It is true that the Spanish had several key advantages the Indians did not, including horse-mounted calvary, firearms, and metal armor. It is also true that European infectious diseases imported by the Spanish, and to which the Indians had not developed immunities, played a large role; within one generation of the Spanish conquest, the population of the Inca Empire shrank to less than a quarter of its pre-conquest size, largely due to disease. Indeed, smallpox entered the Inca Empire before the Spanish did, probably spread from the Spanish explorers in Panama to the Panamanian Indians, who in turn spread it farther during trading voyages to the south. Many scholars argue that disease and armaments were the principal factors – maybe even the sole factors – in determining the outcome of this clash of civilizations.[5] It was also not helpful for the Incas that a royal succession struggle was going on at the time the Spanish arrived, with two sons of the recently deceased Inca Emperor battling for control, and administrative control of the Empire divided between Quito and Cuzco.[6] Although all these factors surely played a role, I, however, am not entirely satisfied with this explanation for why the Spanish prevailed. There must be something more. This is the question that presented itself to me when I walked through the ruins of Macchu Picchu. What was the real determining factor that brought down the Inca Empire? And, what can we learn from the answer?

As in all Inca cities, there is in Macchu Picchu a temple of the sun; at the highest point, there is an observation platform from which observations of solar motions would have been made. So, the Incas had a "science," in so far as observations and measurements were made for the purpose of determining the course of natural phenomena. The information obtained from these measurements was used to establish and control the ritual calendar, which in turn was the instrument of the Inca's control over the population. In other words, the Inca himself had complete and exclusive control over information gathering and dissemination. Only the Inca could make the measurements, knew what measurements to make, what they meant, and how they should be interpreted. There was an entire mythology of divinity built up around this function of the Inca. It was his control of information that gave him his authority and power. It was his control of information that stabilized the Empire.

All hierarchical organizations of society can maintain themselves in only one of two ways, either by coercion or by consent. Maintenance by consent only works when all the members accept their place and role in the organization. This was, in fact, largely how Inca society functioned and remained stable. Although coercion certainly played a role, the key fact is that people regarded themselves not as autonomous individuals, but as existing only in harmony with a society that is watched over by a paternalistic god whose control of information and information flow was as necessary to their survival and wellbeing as was the everyday motion of the sun across the sky. It was as natural to them to accept the authority of the Inca as it was for the child to accept the authority of his or her father. Thus, the entire society was a single organic unit, a family, rather than a confederation of individual and semi-autonomous units, and the integrity of the unit depended on everyone accepting his or her role in it. Such role acceptance implied accepting the exclusive authority of the Inca to measure and interpret natural phenomena.

Even though there was "science" in the Inca Empire, there was no scientific paradigm or scientific methodology or culture. Such concepts had not been invented yet, not even in Europe, where the Renaissance was just beginning in Italy. It would be another 32 years after the Pizarro expedition before Galileo was born. So, the Spanish had no superior ideology, organizing principle, or world view that guaranteed their dominance over the Indians in South America. After all, they also lived in a centrally controlled, hierarchical society presided over by a ruler who also claimed to receive his authority from a divine source. European feudal society was certainly different in many respects from Inca society, but in its key organizing principle – the "divine right" of the ruler to control information and information flow for the benefit of his subjects – the two societies were identical. Rather, the Spanish became lucky beneficiaries of the consequences of the people's acceptance of this basic organizing principle. Once Pizarro killed Atahualpa Inca, establishing in the minds of the people that he was a more powerful god, it was natural for the people to confer upon him the same authority previously held by the Inca. It was as if, as Frances Fitzgerald says in relation to the Confucian societies of Asia,[7] the "mandate of heaven" had changed, and that was all, so it was incumbent upon the people to re-establish themselves in harmony with the new order. Only in that way could the family, the society, survive. Thus, there was no revolution in thinking, no shift in world view, no new paradigm – simply a change of order under the old organizing principle. The Spanish conquest was merely a change in the mandate of heaven.

This is not to say that there was no opposition to the Spanish conquest. Indeed, it was Indian opposition that led to the original search for Vilcabamba by the Spanish. While the Spanish supported a "puppet" Inca government in Cuzco, first headed by Tupa Huallpa Inca, a brother of the murdered Atahualpa Inca, another brother, Manco Inca, retreated into the jungle with a band of followers and set up a parallel government. This government-in-exile continued to wage an on-again, off-again guerrilla campaign against Spanish forces for the next forty years. After the death of Charles I, the new Spanish king, Phillip II, appointed Francisco de Toledo as Viceroy of Peru, and charged him with finishing off Inca opposition once and for all. After arriving in Peru in 1569 Toledo banished the puppet Inca from Cuzco, confiscated and burned thousands of idols, mummies, and *quipus,* resettled rural Indians into Spanish villages to instruct them in proper behavior, and then launched a military campaign to destroy the Inca state in Vilcabamba. Indian opposition to Spanish rule effectively ended when Tupa Amaru Inca was captured in the jungle and executed by Toledo in 1571.

Certainly, the contact of Europe with the Americas was a

cataclysmic historical event that changed the lives and fortunes of people of both cultures. But it did not change the fundamental worldview, or how people thought about organizing their societies and their governments. That change would come more than two centuries later, and on a different continent.

Thus, the Spanish did not learn from their conquest a fundamental lesson. Had they thought rationally about what had happened in the 1500s in South and Central America they could have concluded that Spanish society, and, indeed, all of European society was not invulnerable to the same fate that had befallen the Incas. Indeed, only 16 years after the execution of Tupa Amaru Inca, the Spanish Armada was destroyed by a small English fleet in the English Channel, the beginning of a steady decline in the power and fortunes of the mighty Spanish Empire. It was not in their military preparations and strategies that the Spanish Empire was vulnerable; it was in their unquestioning belief in the fundamental organizing principle of their society. Their fundamental organizing principle was the same as the fundamental organizing principle of Inca society. Had the Spanish recognized this fundamental fact in the sixteenth century then the whole course of history may have been different. But they did not yet have the analytical tools or the ideas that could have led them to this recognition, let alone even conduct the examination. Nor were they motivated to make this examination. To them, the conquest was proof enough of the superiority of the Catholic god and the Spanish king, so there was no need even to question the basic concept of divine right.

Further, the Spanish were very effectively blocked from critical thinking and exploration of new ideas by the force of the Inquisition. The Catholic Church had established an official mechanism to deal with adherence to Church doctrine in response to the Cathar heresy that first arose in southern France in the eleventh century. Heretical ideas imported from Eastern

Europe began to grow even more rapidly in the wake of the First Crusade in 1096, and there was a particularly rapid spread of Catharism after 1140. In 1184 Pope Lucius III and the Holy Roman Emperor Frederick I Barbarossa issued the papal decree *Ad abolendam* in Verona, which established the mechanism for ecclesiastical trials, with those found guilty of heresy in these trials to be handed over to the secular authorities for punishment. The Albigensian Crusade in the early thirteenth century was specifically directed at wiping out the Cathars in southern France.[8] The Holy Office of the Inquisition spread itself throughout Catholic Europe even after the Cathars were destroyed. Both the Dominican and Franciscan Orders were founded to provide the Church with trained inquisitors. The Orders worked with varying degrees of diligence in the struggle against heresy throughout Italy and France, particularly, and also in German-speaking areas of Catholic Europe.

The significant happening in Spain after the defeat of the Moors in 1492 was the co-option of the Inquisition by the Spanish state.[9] The Spanish Inquisition became an arm of official state policy, and was directed by the Spanish Court instead of by the Vatican, as it had been previously. The Spanish Inquisition was largely directed at the Jews. In 1492 Spain contained the world's largest Jewish population.[10] The Inquisition's effect, though, was felt throughout Spanish society. For example, all books and other printed materials not officially approved by the Spanish official who headed the Holy Office in Spain were banned. Agents of the Holy Office were present at all ports and other border crossings to assure that no banned materials came into the country, and that people in possession of such materials would be arrested and duly punished. These agents were, in effect, a police force that controlled the people's thoughts, behaviors, and actions. Just like in Inca society, the head of state, ruling by divine right, controlled all access to information.

The challenge to the concept of divine right would have

to wait until the development of the scientific paradigm, and its application later to the organization of self-government by English-speaking revolutionaries in North America.

1. *We Now Know: Rethinking Cold War History* by John Lewis Gaddis, Oxford University Press, Oxford, U.K., 1997.
2. The term "Inca", although in modern usage coming to refer to the whole of the Indian population of the Inca Empire, was originally used in reference only to the ruler, who was considered to be a direct descent of the Sun god, and to his immediate ethnic clan. Manco Capac Inca, the first mythological Inca, was thought to have emerged from the sacred place of origin of the universe on the Island of the Sun in Lake Titicaca. Divine significance was associated with natural features of the Andean landscape, so Manco Capac Inca's claimed origin from the most significant of these divine places, the origin place of the Sun itself, was meant to justify the Inca's claim to divine right. The descendants of the Indians who inhabited the Inca Empire are now called Quechuas, after the spoken language.
3. Of course, it is not known precisely what the population of the Inca Empire was at the time of Pizzaro's arrival in 1532. Estimates range from 3.5 million to more than 10 million. I have taken the estimate of 10 million from *The Incas and Their Ancestors – The Archaeology of Peru* by Michael E. Mosely (Thames and Hudson Publishers, London, 1992). Mosely says that the ethnic clan of the Inca numbered somewhat less than 40000 people. Thus, most of the people of the Inca Empire were members of subject tribes.
4. An excellent account of the details of Inca astronomy can be found in *Astronomy and Empire in the Ancient Andes* by Brian S. Bauer and David S. P. Dearborn, University of Texas Press, Austin (1995).
5. See, for example, *Guns, Germs, and Steel* by Jared Diamond,

W. W. Norton and Co., New York (1997).

6. There are many excellent books on Inca culture, religion, and history, and on the Spanish conquest. One particularly good source is "The Incredible Incas and Their Timeless Land", by Loren McIntyre, National Geographic Society, Washington, D. C. (1975).
7. *Fire in the Lake* by Frances Fitzgerald, Little, Brown and Co., 1972.
8. *Montaillou: Cathars and Catholics in a French Village 1294-1324* by Emmanuel LeRoy Ladurie, translated from the French by Barbara Bray (Scolar Press, London, 1978).
9. *The Inquisition, Hammer of Heresy* by Edward Burman, Dorset Press, New York, 1992.
10. In the aftermath of the Roman-Jewish War and the destruction of the Jewish kingdom of Judea and its capital city, Jerusalem (the Roman victors first burned down the city and then turned it into a Roman Army camp), the surviving Jewish population was largely dispersed. Some people migrated north, ending up in areas of Eastern Europe and becoming the ancestors of the modern-day Ashkenazic Jews. The larger number migrated south, into areas that later came under the control of an expanding Moslem empire that spread across northern Africa and then into Spain with the Moors. Moorish Spain became a center of a flourishing and vital Jewish and Moslem culture. Moorish control of the Iberian Peninsula was lost to the then newly created Kingdom of Spain in 1492, and then both Moorish and Jewish culture was effectively destroyed by the Inquisition in the aftermath of the Spanish conquest. An excellent account of Jewish history can be found in *A History of the Jews* by Abram Leon Sachar (Alfred A. Knopf Publishers, New York, 1974). There were a number of "secret" Spanish Jews who escaped the Inquisition in Spain by joining the Spanish voyages of exploration to the New

World. Not only Jews came to the New World from Spain, however. So did the Inquisition. The Headquarters of the Spanish Inquisition in the New World was established in Lima, the Spanish capital of Peru.

Chapter 2

The European Enlightenment Comes to America

"Almost everything that distinguishes the modern world from earlier centuries is attributable to science, which achieved its most spectacular triumphs in the seventeenth century." These words were written by the mathematician-philosopher Bertrand Russell in his comprehensive work *A History of Western Philosophy*.[1] Although Russell's statement seems at first glance overdrawn and difficult to justify, it is not, for the simple fact that the seventeenth-century scientists not only made important discoveries about the natural world, but, more importantly, changed people's whole way of thinking about the natural world and their place in it. These new conceptions in science came to be linked to other radically new conceptions about the organization of society and government.

Just seven years before Tupac Amaru Inca was murdered by the forces of the Viceroy of Spain in Peru, Galileo Galilei was born in Tuscany. It was Galileo who, building on the traditions established earlier by Copernicus, Tycho, and Kepler, of using careful observation and measurement, along with the principles of inductive reasoning, ascertained the basic laws of motion. His science of mechanics formed the foundations of modern physics, and his dependence on experiment to test hypotheses remains to this day the cornerstone of the scientific method.

His experimental discoveries about motion had a profound effect on changing our conceptions about how the world works. Prior to Galileo, most people accepted the theological view that the earth stood still in the center of the universe, and that the heavenly bodies, forever immutable, moved in "natural" circular motions around the earth. Since only living beings possessed the property of motion in the world observed by people, it was thought that the moving heavenly bodies were themselves "living" deities (as the Incas believed), or that (as Europeans believed) they were moved by divine will. The regularity in the motion of heavenly bodies was taken as a clear manifestation of the immutable Will of God. Thus, circular motion was thought to be natural motion for all celestial objects. One of Galileo's greatest contributions to human thought was his demonstration that motion in a straight line, not circular motion, is natural motion. Thus, without doing something to change the direction of a moving object, it would naturally continue to move in a straight line. He also introduced the concept of acceleration, that is, the change in the direction or speed of a moving object, and with his famous experiment conducted from the top of the leaning Tower of Pisa, showed that the acceleration of gravity is a constant, independent of an object's size and mass. Prior to this experiment, it was assumed that a lead cannonball, for example, would fall faster than a lead bullet. Galileo proved by experiment that this old assumption is false.

Galileo also established the laws governing motion of projectiles. Perhaps his most important finding was that such motion could be described as a combination of a constant acceleration downward and a constant velocity horizontally. This insight was important in opening the way for a mathematical description of motion, made possible by the development of coordinate or analytic geometry by the French mathematician-philosopher Rene Descartes, who was a contemporary of Galileo, and the development of differential and integral calculus independently by Leibnitz in Germany and Newton in England.

Isaac Newton, born in the same year that Galileo died, provided the final synthesis of the new science of mechanics. To the concept of acceleration, he added the new concept of "force," which he defined as the agency that causes changes of motion. Thus, acceleration, which is a change in motion – either speed or direction or both – is caused by a force. Force is what we humans experience in our everyday lives as a "push" or a "pull." If no force acts, a body in motion continues indefinitely in a straight-line path at constant speed, and thus undergoes no acceleration. If the object is at rest to begin with it can only be put into motion by application of a force. Newton showed that it is the force of gravity, arising in matter itself, that causes the acceleration of heavenly bodies in their orbits as well as the acceleration of the apple falling from the tree. More important, Newton showed that all motions caused by the force of gravity could be explained by the same formula, namely, that acceleration varies inversely as the square of the distance between the moving bodies. Thus, the acceleration of Mars in its orbit around the sun is less than that of the earth, since Mars is farther away from the sun than is the earth. Likewise, all objects at or near the surface of the earth have the same acceleration towards the center of the earth, which depends only on the constant mass of the earth.

Newton's laws of motion were a radical departure from all previous thinking. No longer was divine will required to animate the motion of celestial objects. Once set in motion Newton's law of gravity was sufficient to explain their motions for all future time. People could still believe that God had set the original motion at the time of creation – Newton himself apparently believed that the planets had been "hurled by the hand of God" – but no longer was God required to keep them in motion. Indeed, even the idea that God was necessary to set the original motion of the planets began to fade in the next century when the French scientist-mathematician Pierre Laplace published his nebular hypothesis for planetary formation. In this hypothesis, the planets are thought to have formed out of the gravitational collapse of a large gas cloud, or nebula, orbiting the young sun. Laplace's hypothesis is deduced directly from Newton's laws of motion, and has gained widespread support in the past century as a result of observations made with modern telescopes.

Indeed, the discoveries opened up by the invention of the telescope of previously unknown heavenly bodies (Galileo used this new invention to first sight the four largest moons of Jupiter), and the discovery that the Milky Way nebula is in fact a huge collection of individual stars at great distance, also challenged the prevailing conception that God had created the universe with us at the center of it. Thus, the notion that all of creation had human beings as its purpose became irrelevant to the advancement of our understanding about creation. No longer did people have to accept on ecclesiastical authority the facts about God's creation and their place in it; they could now follow the scientific method and determine the facts for themselves.

It was a natural extension of the new scientific paradigm that people who no longer needed to believe that God moved the planets also no longer needed to believe that their rulers ruled by divine right. The concept of democratic government, however, did not grow directly out of the scientific paradigm. These two ideas of the European enlightenment were not directly linked until they came to America.

In Europe, there were many attacks on the concept of divine right even before Newton's publication of *Principia Mathematica* in 1687. One of the most important of the critiques of hereditary rule was published by the English philosopher John Locke in the first of his two *Treatises on Government* only two years after Newton's *Principia* appeared. Locke's work was done during the time of great political ferment of the Civil War in England, during which King Charles I was deposed and killed and the country faced its first experiment with parliamentary democracy. Ideas of democratic government, however, were not confined to England. Ideas similar to Locke's were being openly discussed and published on the Continent as well.

It was with the founders of the new American republic less than a century later, however, that the linkage of science with democracy was first made explicit. The opening sentence of the Declaration of Independence makes this linkage:

> *When in the Course of human events, it becomes necessary for one people to dissolve the political bands which have connected them with another, and to assume among the powers of the earth, the separate and equal station to which the Laws of Nature and of Nature's God entitle them, a decent respect to the opinions of mankind requires that they should declare the causes which impel them to the separation.*

Note that it is the "Laws of Nature" that entitle the people to self-government, not the Will of God. If there is any doubt that Thomas Jefferson, the author of these lines, was informed by the scientific paradigm in forming his views on democratic government, we have only to look at the rest of what he wrote and said and did. We will have a look at this in a moment. First, however, let us look at the second sentence of the Declaration of Independence:

> *We hold these truths to be self-evident, that all men are created equal, that they are endowed by their Creator with certain unalienable Rights, that among these are Life, Liberty, and the pursuit of Happiness.*

Note that the stated truths are *self-evident,* not revealed by God or by any other higher authority. Facts that are self-evident can be discovered by individuals through the agency of their own reason, and thus not have to await to be known by instruction from the Inca or the King or the Bishop. The equality of individuals is taken as a given axiom, from which the postulates of their government can then be deduced. It is no accident that the language is that of Euclidean mathematics. The logic of the democratic experiment is thus brought full circle to its origin in classical rationalism. The scientific paradigm was all that was needed to complete the circle. Thus, the linkage is made between rational thought, the scientific paradigm, and liberty and self-government. Neither the idea of democratic government nor the ideas of the "Laws of Nature" were original ideas with America's Founding Fathers, but their linkage was. This linkage was a truly radical notion that forever changed human life and the course of human history.

As I will argue in this monograph, it is precisely this linkage that two centuries later ultimately led to the end of the Cold War and the collapse of the Soviet Union.

According to Russell,[2] it was Benjamin Franklin who changed Jefferson's original draft of the Declaration of Independence to include the words "self-evident" in the second sentence. This change, as we have seen, was significant, but also not surprising. Franklin himself represented the very fusion of the liberal democrat and the scientist in his own life and being. The key role he played in the establishment of the new American republic is well known and well documented, and will not be repeated here. His role as America's "first physicist," however, is less well known, and merits some discussion, because his work as a scientist was integral to how he approached the problems of government, and became the model from which modern scientific culture developed.

Franklin understood early on that the advancement of science was integrally linked to a democratic culture that guaranteed freedom of thought, freedom of discussion, and freedom to experiment with unorthodox ideas. His own scientific work was unorthodox, and probably would have been difficult if not impossible in a culture that suppressed unorthodox views. Where Galileo had performed experiments to unravel the nature of motion, Franklin performed experiments to unravel the nature of electricity. His kite-flying experiment, in which he discovered that atmospheric lightning is an electric current flow, has become part of American lore. It is taken for granted now that everyone knows that lightning strikes are the principal cause of naturally occurring forest fires, but prior to Franklin no one had even thought of making the observations or doing the experiments to discover this connection, let alone to determine the nature of lightning. The peoples of rationalist and democratic ancient Greece surely had seen lightning, but if they thought about its nature at all they never thought to test their hypotheses about it. In fact, the cause of natural forest fires was *assumed* to be something else entirely. As Thucydides reports in his *History of the Peloponnesian War*,[3] a contemporary account which he claims to have based on eyewitness reports, "great forest fires – have broken out spontaneously through the branches of trees being rubbed together by the wind." We now know, of course, that such a claim is nonsense, and we are left to wonder how it is that he came up with such a claim. It certainly was not by observation or experiment. It took about 21 centuries after Thucydides before Franklin's experiment provided the factual foundation to test and challenge what everyone, on the authority of Thucydides' claim, assumed to be true.

Less well known than the kite experiment are Franklin's many laboratory experiments on the nature of electric charges,[4] in which he made the important discovery that there are both "positive" and "negative" electric charges that behave in opposite ways. He also discovered the Law of Conservation of Charge. These discoveries were a breakthrough in understanding that laid the experimental foundation for the laws of electromagnetism that were to follow in the next century.

It was Franklin's scientific work that proved instrumental in the success of his diplomatic mission to France to secure French support for the new American republic. His *Experiments and Observations on Electricity* had been published in London in 1769 and widely distributed, so by the time he arrived in Europe on his first diplomatic mission his reputation and fame as a scientist had already preceded him. He was welcomed in Paris primarily as a scientist, was already one of the very few foreign members of the French Academy of Sciences, and was invited to give lectures and demonstrations on his scientific ideas. It is not possible to overstate the effect on America of the high regard the French had for Franklin as a scientist, because Franklin clearly traded on his reputation to secure French financing and military help for the newly independent American nation. French assistance was instrumental in securing an American victory in the Revolution.

More important, Franklin's scientific work helped to cement the idea that science and democracy are integrally linked into a "single cultural mode," because Franklin's scientific work was possible only within the context of the emerging American democracy. This single cultural mode, as defined by David Hollinger,[5] means "that the scientific enterprise was an expression of democratic political culture, and that the autonomy of science depended upon the strength of democracy."

Although not himself a scientist, Thomas Paine certainly shared this view. Paine, the pamphleteer of the Revolution, first came to America from England in 1774 at the age of 37 with a letter of introduction from Franklin; the two men had met there and formed an instant bond of friendship that was to last for many years. It was Paine, in his first publication *Common Sense,* who argued for independence, a complete separation from England. It was also Paine who cheered on the Revolution and the democratic experiment through his series of papers entitled *The American Crisis*. General Washington read one of these papers to his troops to help lift their spirits during the dark days of the Continental Army's winter at Valley Forge.

Later, President Washington, the Federalist, split with Paine over the issue of American support for the French Revolution.

Paine became a leading advocate for the overthrow of the monarchy in France (although he argued, unsuccessfully, against the execution of King Louis XVI).[6] In fact, he went to France after the outbreak of the French Revolution, became a French citizen and an elected member of the National Convention, assembled to draft a constitution for the newest democratic republic. After he published a written attack on Jean-Paul Marat, leader of the Jacobin faction in the Convention, he and several other members of the Girondin faction were arrested in 1793 and Paine spent a couple of years in prison. He was one of the few to avoid the guillotine during the Terror, but what he saw as President Washington's failure to intercede with the French government on his behalf permanently alienated him from Washington and the Federalists.

By this time, however, he was going in a different direction than the Federalists with his ideas on democratic government and scientific rationalism. His publication *Rights of Man* was a spirited rebuttal of Edmund Burke's attacks on the French Revolution; Burke had previously put himself at risk in his own country by being one of the few Englishmen to openly support the American revolutionaries. The French Revolution, on the other hand, was too much for an Englishman, even Burke, to bear. Paine based his rebuttal on what he viewed as people's natural rights, sentiments very similar to those expressed in the American Declaration of Independence. These ideas, in combination with his ideas on the scientific paradigm expressed in his *Age of Reason* provided the most explicit linkage of the time of science with democracy. Paine wrote *Age of Reason* as an assault on the authority of the Christian bible. His thesis was that man can know God and His wisdom through science, not by simply accepting "truths" of the bible that are demonstrably untrue. Only by the "divine gift of reason that God has given to man" can come the "knowledge that man gains of the power and wisdom of God." Scientific principles can be employed to make predictions about future natural occurrences, like eclipses, and these predictions can be tested against observation to reveal the majesty of a God that created a universe governed not by arbitrary will but by natural law. "We can know God only through his works – The principle of science leads to this knowledge."

Jefferson had read Paine's writings and was strongly influenced by them. He, too, believed that democracy and rationalism were necessary for the advancement of science. "It (democracy) is the great parent of science & virtue: and that both, always in proportion as it is free," he wrote.[7] Jefferson had also read Locke and Newton. Indeed, he thought that "Bacon, Locke and Newton" were "the three greatest men that have ever lived, without any exception."[8] Not only were the complete works of Locke in Jefferson's extensive personal library but so were the complete works of the French political philosopher Montesquieu as well as books by other English and French thinkers.

Jefferson also believed that scientific advancement provided a net benefit to mankind and to the prosperity of the United States. Although he convinced Congress in 1803 to fund the Lewis and Clark expedition "for the purpose of extending the external commerce of the United States," he was clearly also very much interested in the new plant and animal discoveries that such an expedition was bound to make. Jefferson placed great value on the many objects the expedition leaders sent back (many trunks full of new plant species, animals, and minerals; there were even cages of live animals), and especially on their diaries that in a very real sense were like a scientific report on a scientific field trip. The expedition also advanced the sciences of cartography and linguistics. Jefferson took a special interest in doing a comparative study of Indian languages, the first such study ever undertaken. It could even be said that he originated the field of comparative linguistics. The third President also promoted and advanced technological innovation, both by his policies and by his own personal example. In designing and building his Monticello plantation manor house he became a self-taught architect who adapted classical architectural concepts to new building technologies, and he invented several innovative "modern" conveniences for his house, including the dumb-waiter, a revolving desk, and a mechanical copying machine (which he called a polygraph). He also did the architectural design of the University of Virginia campus in nearby Charlottesville.

Jefferson's rationalism, however, provoked substantial opposition (as did Paine's *Age of Reason*) from conservative religious believers who felt that the social order and ethical behavior that depended, in their view, on religious belief was under assault from the "evil" forces of rationalism and science. Rationalism, in their view, could only lead to war and chaos and immorality. It was, ironically, the French Revolution that confirmed their anti-rationalist views and first split the American people into two opposing political parties largely based on whether they supported or opposed the French. The Republican-Democrats, led by Jefferson, were firmly pro-French; they had the radical view, as Paine had said, that "the government that governs best governs least," a sentiment that came to justify the excesses of the radical Jacobins and the Reign of Terror. In the view of the eighteenth-century Democrats, if one accepts the rationalist or scientific paradigm, then any and all challenges to authoritarian government become inevitable. Both Jefferson and Paine were unapologetic supporters of the French revolutionaries and refrained from criticizing the French Revolution's degeneration into an orgy of violence.

To the Federalists, on the other hand, the violence and chaos unleashed by the French Revolution, and the outbreaks of war in Europe that followed in its wake, were precisely what was wrong with the rationalist view. People who believed that strong central government based on religious principles was necessary to maintain the strength, cohesion, and social order of the new republic identified themselves with the Federalists. Further, the Federalists believed that prosperity for the new republic rested on maintaining strong trade ties with the recently vanquished British, which implied strict neutrality on the part of the United States and non-involvement in European wars and affairs.

The French Revolution put this first American political conflict into sharp perspective. The Federalists had only to point to Europe to argue their point that rationalism, which was the guiding principle of the French Revolution, only leads to disaster. The Republican-Democrats, on the other hand, saw the principles of American democracy shaking the very foundations of the old order in Europe, which they viewed as a positive development that the United States should support. This was indeed the main issue in the Presidential election of 1800, which Jefferson of course won. Indeed, Jefferson himself recognized the significance of his victory in the 1800 Presidential election, because it was the first time in the new republic's history that governing power was transferred peacefully to an opposition political party. Jefferson later called this event the Second American Revolution.

The American Revolution and the French Revolution were both revolts against the principle of rule by divine right and an affirmation of the *self-evident* right of people to govern themselves. The different courses these two democratic revolutions followed, however, exposed some internal contradictions in the idea of democratic rule, for if rule by divine right is to be replaced by the rule of law, what or whose law are we talking about? To the Jeffersonians and the Jacobins this meant the "Laws of Nature," as stated in the Declaration of Independence; since any individual following the precepts of the rationalist or scientific paradigm could figure out for himself or herself what these natural laws are, a strong central governing authority

– whether King or Church – was not only unnecessary but to be resisted, even with violence. In France, where there was no tradition of religious plurality as in England and America, this view became an enforced uniformity of ideology. In America, with its multiplicity of religious groups, there was a clearer recognition of the need for a strong central governing authority, not to enforce a particular governing ideology but to maintain the social harmony between the different groups necessary to protect private property rights, promote prosperity, and protect religious freedom. Jefferson himself even recognized this, and worked hard to guarantee that the new democratic experiment would not be subverted by one group establishing supremacy over any other. He most feared the establishment of a state church, or state religion, which he was sure would lead America into the same quagmire of religious wars and tyranny that he viewed as plaguing Europe. Jefferson considered his greatest accomplishment to be not the Declaration of Independence but the implementation of the idea of separation of state and church that was first embodied in the Virginia Charter and later in the Constitution of the United States.

Jefferson's victory in the Presidential election of 1800 did not settle the conflicting issues for American democracy raised in that election. Some variant of the rationalist-religious conflict continues to play itself out every four years in the United States. These contradictory currents, which are still present in contemporary American society, have exposed some unintended consequences of the linkage of the scientific paradigm with liberty and democracy. The same democratic government that protects free thought so necessary to rational science also protects non-rational beliefs leading to fanaticism and zealotry. Democratic government's protection of individual rights also has implied in the United States the protection of private property rights leading to support of slavery until the middle of the nineteenth century, concentrations of economic power, and weakening of government's ability to guarantee social harmony. It is precisely the last point that had the old Federalists so worried. Their old worry has been taken up by modern-day conservatives, while worries about unrestrained economic power deriving from private property rights consume modern-day liberals. They would both probably be surprised to learn that their worries stem from the same source. They would also both probably be surprised to learn that the resolution of the contradictions might be able to be derived from a further linkage that was first made in the twentieth-century Soviet Union.

1. *A History of Western Philosophy* by Bertrand Russell, Simon and Schuster, New York, 1963.
2. Russell does not provide a source for this information, so it is not known how he knew that it was Franklin who made this change in Jefferson's original draft. The document that has since become known as the "Original Draft" that was submitted to the Continental Congress from the Committee of five people appointed by the Congress to produce a draft Declaration of Independence was Jefferson's original draft with hand-written changes on it, but with no attribution of who was responsible for which change. In later years, when Jefferson became concerned about his legacy, he claimed repeatedly that the Committee made no substantive changes to his original draft, implying that all the hand-written changes were his own. He also expressed much criticism of the Congress for removing from the list of grievances against King George in the Original Draft the statements on the King's responsibility for the slave trade. Jefferson's attitudes and views toward slaves and slavery were complex and contradictory, the subject of much recent writing, and will not be discussed here.
3. *History of the Peloponnesian War* by Thucydides, translated

by Rex Warner, Penguin Books, London, 1954.

4. Franklin reported on his experiments on electricity and electric charges in letters written to his friend Peter Collinson, a Fellow of the Royal Society in England. These letters were collected and published in London in a book entitled *Experiments and Observations on Electricity Made at Philadelphia in America* in 1769.
5. *Science, Jews, and Secular Culture* by David A. Hollinger, Princeton University Press, Princeton, New Jersey, 1996.
6. See *Collected Writings of Thomas Paine,* edited by Eric Foner, the Library of America, New York, 1995.
7. This sentence is contained in a letter by Jefferson to Joseph Willard, March 24, 1789; the original letter is available in the Library of Congress, Washington, D. C.
8. This sentence is contained in a letter by Jefferson to John Trumbull, February 15, 1789; the original letter is available in the Library of Congress, Washington, D. C.

Chapter 3

The European Enlightenment Comes to Russia

The Russian Revolution[1] was, like both the American and French revolutions, a cataclysmic rejection of rule by divine right. Unlike in America and France, however, the new Russian state was, from its very inception, a totalitarian state guided by an ideology unlinked entirely from the principles of the European enlightenment that motivated the other two revolutions. Instead of turning subjects into citizens, equal and free under the rule of law, the Russian Revolution turned subjects into comrades, working in unison to advance classless socialism under the rule of the Communist Party. As such the new state could only maintain its power by suppressing the unfettered interchange of ideas and free inquiry on which the progress of science depends. It was therefore inevitable that it would be Soviet scientists, and their attempts to maintain linkages with the wider world of international science, that would ultimately provide the intellectual force that brought down the Soviet state and ended the Cold War.

It was not the scientists, however, who started the Soviet dissident movement. It was the poets. They, in turn, were emboldened to speak out first by the death of Stalin in 1953 and then by Khrushchev's denunciation of "the cult of personality" at the Twentieth Party Congress in February 1956. Joseph Stalin, who had seized control of the Soviet Communist Party and the state governing apparatus after the death of Vladimir Lenin in 1923, had waged a relentless 30-year campaign of terror that had murdered all political and cultural opposition and potential opposition to the regime. Estimates of the number of people killed by the Soviet regime during the period of Stalin's forced collectivization of agriculture range up to more than 20 million. At the time of his death in 1953, there were several million political prisoners in Siberian prison camps, people who had simply disappeared from their homes without a single protest being raised. Khrushchev's decision to begin the release of these political prisoners, and their return to the cities and towns of their residence, unleashed changes that the regime, to its horror, found itself unable to control.

The first was the challenge by the literary community to state literary censorship. Soviet writers could publish only if they were members of the officially sanctioned Writers Union. Boris Pasternak's decision to send the manuscript of his novel *Doctor Zhivago* to Italy after it was rejected by Soviet publishers was seen by the Soviet government as a direct assault on their absolute control of all contact with the West. No Soviet writer before Pasternak was bold enough to challenge the government's control of literature in this way. After Pasternak was awarded the Nobel Prize for literature and refused at first to accede to Soviet government demands that he refuse the award, he was expelled from the Writers Union.

The struggle for literary freedom took a dramatic turn just a year and half after Khrushchev's Party Congress speech, and less than three months before the Nobel Prize award to Pasternak. Then, on July 29, 1958, an official ceremony was held in Mayakovsky Square in Moscow for the unveiling of a statue of the poet Vladimir Mayakovsky.[2] Some official Soviet poems were read at the conclusion of the ceremony, after which some people in the audience began spontaneously to read their own uncensored poetry. This episode evolved quickly into regular monthly gatherings in Mayakovsky Square for "illegal" poetry readings. The government, however, was not going to allow these affronts to "socialist solidarity" to go unmolested. Police harassment and arrests were increased, with the intent of stopping the Mayakovsky Square gatherings. As the repression increased, however, so did the popularity of the gatherings, which began to attract artists, mathematicians, and a wide array of the Moscow intelligentsia. Every arrest and imprisonment that stemmed from the Mayakovsky Square gatherings just led to a further escalation of dissident activity, as the protesters sought to fight back by clandestinely printing and circulating accounts of the arrests, the trials, and the imprisonments, along with the "illegal" poems and essays and stories.

Open warfare between the government and the literary community erupted with the arrest of the Soviet writers Yuli Daniel and Andrei Sinyavsky in September 1965. By this time Khrushchev had been deposed as Soviet Premier, to be succeeded by Leonid Brezhnev. Sinyavsky was no ordinary young poet-protester from the Mayakovsky Square gatherings. He was a well-respected senior member of the Writers Union, a senior staff member of the Gorky Institute of World Literature, and a scholar and teacher at Moscow State University. His arrest sent shock waves through the Moscow cultural community. What was most astonishing is that the principal evidence used against Daniel and Sinyavsky at their trial was their own officially published work.

By early 1966 rumors were circulating that Brezhnev would rehabilitate Stalin at the next Communist Party Congress in March. There was genuine alarm in the intellectual communities all over the Soviet Union at this prospect. Signatures were gathered on three petitions that were sent to Soviet authorities. Two of the petitions were directed at seeking redress for Daniel and Sinyavsky. The third petition warned that it would be "a great disaster" for the Soviet Union if the Communist Party were to rehabilitate Stalin. What was most remarkable about this petition, however, is that in addition to signatures of prominent members of the literary and artistic communities it also contained the signatures of the three most prominent physicists in the Soviet Union. Igor Tamm had already received the Nobel Prize in Physics. Pyotr Kapitsa would win it just twelve years later. The third physicist to sign the petition was Andrei Sakharov, who had earlier created the designs for Soviet thermonuclear weapons, and had more recently become an outspoken advocate for arms control. This was his first public protest. Afterwards, neither he nor the Soviet dissident movement would ever be the same. In fact, his role in the events that were to unfold over the next two decades is so pivotal that the next chapter is devoted entirely to a discussion of his work and ideas.

The scientists' petition to Brezhnev transformed what had been simply an attempt by a few eccentric poets to read their poetry in public into a major dissident movement.[3] The dissident movement in the Soviet Union was created, largely by the scientists, out of the stitching together of several independent threads. Sakharov and the other scientists provided not only the intellectual linkages to the ideas of the European enlightenment, but also the personal linkages to what had been up to that time several other independent currents of dissent in the Soviet Union.

One such current was the Soviet Jewish emigration movement. Another was the human rights movement. A third was the peace, disarmament and arms control movement. The three movements initially had quite separate goals and objectives, not to mention different strategies. It was Sakharov and other Soviet scientists who were the first to recognize the linkages between these seemingly disparate groups and who first brought them together.

Soviet Jews, victims of chronic Russian antisemitism that predates the Revolution, began a period of cultural and spiritual renewal at the time of the Khrushchev "thaw" that was greatly accelerated after the Israeli victory in the Six-Day War in the summer of 1967. This cultural and spiritual reawakening was suppressed at every turn by the state. Books printed in Hebrew and Yiddish were banned. There were very few synagogues, with state-appointed rabbis, to serve what was then the world's second-largest Jewish population, some four million people. Jews were identified as a separate nationality group on the internal passports that all Soviet citizens were required to carry, and this identification often left them exposed to discriminatory behavior in all aspects of their state-controlled lives, from housing to education to employment. Worse, Soviet propaganda tended to identify Jews with Zionist enemies of socialism. Thus, Soviet Jews, particularly those who sought to exercise their cultural or spiritual identification, were looked upon as potential traitors or enemies by the rest of the society. As a result, Soviet Jews sought emigration as their escape from the increasingly untenable situation in which they found themselves. Unlike the human rights activists, they were not seeking to reform the regime. They were simply seeking to escape from it. Since emigration was not a legal right for Soviet citizens, the Jewish emigration movement was born. Its objective was to liberalize and open up the emigration laws to allow Jews to "repatriate" to their natural homeland in Israel. Those Jews who applied to emigrate and were refused an exit visa, as almost everyone was, became known as "refuseniks."

Some, however, felt compelled to take more extreme measures to get out. Twelve people were arrested by the Soviet secret police, the KGB, at Smolny Airport in Leningrad on June 15, 1970, and accused of conspiring to hijack an airplane. Even though the "conspirators" had not even approached an airplane, the response of the Soviet government to this alleged hijacking plot was immediate and harsh, with the intention of crushing once and for all the Jewish emigration movement before it could even get off the ground. In a trial barred to foreign observers, the leader of the alleged plot, Edward Kuznetsov, a veteran of the illegal poetry readings in Moscow's Mayakovsky Square in the days before the Sinyavsky and Daniel trial, received the death sentence along with Jewish activist Mark Dymshits. Only massive public outcry from all over the world prevented the implementation of these sentences. The public response, in turn, came about because human rights and democratic activists in the Soviet Union had devised a scheme to get information about the Leningrad trial out of the courtroom and into the hands of foreign reporters in Moscow. When the Supreme Court of the Russian Republic heard the appeal of the death sentences and commuted them to fifteen years in a labor camp, it was Sakharov who announced the judgment to the crowd assembled outside the court building.

The Leningrad trial, and Sakharov's involvement, began the process of change in the relationship between the human rights and Jewish emigration movements. Some Jewish activists had been involved in the human rights movement, but many more were not, and some were very much opposed to involvement with the democratic activists because they believed strongly that such activity would hamper their chances for emigration. On the other side, many of the democratic activists were repelled by the provocative tactics and inflammatory rhetoric of the Jewish emigration activists; the democrats were attempting to liberalize the system, not antagonize it. It took Sakharov and a few others to recognize that, although the two movements had different goals and tactics, they were really linked because the freedom to choose one's country of residence is a fundamental human right. Thus, the democrats of necessity were invested in the success of the Jewish emigration movement, and the Jewish emigration movement also depended on its success on the success of the democrats in getting the Soviet government to obey its own and international laws. It was the scientists who recognized this linkage because they already understood the linkage between science and democracy. Thus, it is not surprising that it was largely the scientists who became active in both movements simultaneously, whether they were Jewish or not (Sakharov was not), and who provided the intellectual underpinnings of the linkage of the two movements.

The focus for the linked movement, then, turned to exposing the rational contradiction that existed between the Soviet government's opposition to linkage and their legal obligations under Soviet and international law. The contradiction was that not only did the constitution of the U.S.S.R. and the U.N. Declaration of Human Rights, to which the Soviet Union was a signatory, guarantee by law the same rights, including the right of emigration, for which the activists were fighting; the U.S.S.R. was also signatory, along with 35 other nations, of the Helsinki Accords that formally incorporated the concept and practice of linkage as part of international law. These Accords, signed in 1974 at the conclusion of the Conference on Security and Cooperation in Europe (CSCE) held in Helsinki, Finland, consisted of three general parts. Basket I of the Accords formalized the political *status quo* in Europe, and was seen as a significant victory in the U.S.S.R. for Brezhnev's policy of *détente*. It also, more importantly, delineated agreement on a number of "confidence building measures" to lessen the chances of armed conflict between the NATO and Warsaw Pact forces. These measures included, for example, the requirement for each side to notify the other in advance of any large troop movements or military training maneuvers, and a ban on encrypting missile test data telemetered to ground stations during test flights. As such, Basket I of the Helsinki Accords was, in many respects, an important arms control treaty. Basket II of the Accords called for promotion and expansion of contacts, communications, and information exchanges in the fields of economics and business, science and technology, and the environment. Basket III pledged each signatory to guarantee to its citizens all those basic human rights already delineated in both the U.S. and U.S.S.R. constitutions. These included freedom of religious expression, freedom of the press, freedom of movement within one's own borders, and freedom of emigration. Thus, the Helsinki Accords, by their very structure, codified linkage into international law.

The most important effect of the Helsinki Accords in the Soviet Union was the formation in every major Soviet city of Helsinki Watch Groups to monitor their government's compliance with the treaty. These groups formalized the link that already existed between the democratic activists and the Jewish emigration activists. It was largely Soviet scientists who constituted the membership of these groups. The group activists argued that since the record of Soviet compliance with Basket III of the Accords was very poor, despite Soviet protestations to the contrary, the Western countries were naive to expect Soviet compliance with Basket I, which is considerably more difficult to monitor.

The Soviet government, in their turn, saw these groups as a major threat since they called into question, through linkage, Soviet compliance with arms control agreements. The Helsinki Watch Groups thus gave the lie to the Soviet contention that it was militarists and "imperialists" in control of Western governments who were blocking the U.S.S.R.'s sincere efforts at serious arms control. Instead, the possibility that the Soviet government could not be trusted to comply with arms control agreements was coming to be recognized as a major block to such agreements. The Soviets, therefore, had to silence their internal critics lest such criticism destroy their international credibility. Silence them they did. Of the original two dozen or so founding members of the Moscow Helsinki Watch Group, only one, Yuri Yarim-Agaev, managed to emigrate early on. All the others, except for two, were sent to prison, labor camp, or internal exile. These included prominent democrats, like physicists Andrei Sakharov and Yuri Orlov, and leaders of the Jewish emigration movement, like computer scientist Anatoly Shcharansky.

It was not enough, however, to silence the democrats and

Jewish activists. The Soviet government also sought to destroy the credibility of their arguments by calling into question their mental competencies and their loyalties to socialism and the U.S.S.R. Hence the statement by Vitaly Ruben, a senior Soviet official, at a press conference in Moscow in December 1983 that Sakharov was sent into internal exile "first of all for his own piece of mind."[4] Hence also the escalation in the campaign in the Soviet press to identify Soviet Jews as part of a worldwide Zionist conspiracy intent on destroying socialism. Individuals were labeled by name in Soviet newspapers as traitors to the state. Such statements provoked several incidents of anti-Jewish violence in several cities, most particularly in Moscow and Kiev.

Another development in the Soviet campaign to discredit the dissidents was the formation in April 1983 of the Anti-Zionist Committee of the Soviet People,[5] ostensibly organized to give official sanction to the "popular" struggle against the "evils" of Zionism, but in fact, serving as a justification for antisemitic activities of all sorts. This included a resurrection of that notorious Czarist forgery *Protocols of the Elders of Zion* as required reading for many Soviet schoolchildren, and the crude misuse of Jewish religious symbols for anti-Zionist propaganda purposes.[6] In a situation hauntingly reminiscent of Nazi Germany of the 1930s, Soviet Jews found their way increasingly blocked by a plethora of restrictions to entry into the major universities and to upper-level positions in the civil and military services.[7] The internal passport regulations were increasingly used to bar their residence in certain areas and as a pretext to their imprisonment.[8] There was virtually no area of Soviet life – work, education, housing, army service, public services – that was free of an all-pervasive antisemitism that was officially sanctioned and encouraged by the government.

These practices began to affect not just the democratic activists and the Jewish emigration activists, but ordinary citizens as well. The plight of refuseniks – Jews who applied to emigrate and were refused – was even worse. Typically, a refusenik lost his or her job as a result of an emigration application, and since it was a criminal offense in the U.S.S.R. to be unemployed the refusenik was then subject to prosecution and imprisonment for "parasitism," as well as for a bewildering array of more serious charges, being now identified as a Zionist enemy of socialism. Despite the risks, approximately half the entire Jewish population of the Soviet Union sought to emigrate during the 1970s and 1980s. These were the people who were the intended targets of the Anti-Zionist Committee's work.

Additionally, the Soviets tried to lend credibility to the Anti-Zionist Committee's work by constituting it with the country's most prominent assimilated Jews. The Chairman of the Committee was Colonel-General David Dragunsky, a much-decorated Soviet Jewish tank commander who at the time of the Committee's formation was head of a military training institute near Moscow. It was Dragunsky's institute, according to documents captured by the Israeli army in Lebanon in 1982, at which leaders of the PLO guerrillas were trained.

The Committee's Vice Chairman was Samuel Zivs, a prominent Soviet jurist and propagandist. It was he who appeared at a press conference in Moscow on June 6, 1983, to announce to the world that Soviet Jewish emigration was ending because there were no more family reunifications to be accomplished,[9] despite the large number of refuseniks who were then awaiting permission to join family members already in Israel or the West.

The Anti-Zionist Committee was used to discredit the Jewish emigration and cultural movements in the U.S.S.R. It was also used to discredit linkage. Thus, in a broadcast by the North American Service of Radio Moscow on November 10, 1983, a group of Soviet assimilated Jews appealed to Jews of the United States to disregard "the myth of a Soviet threat." They went on to say that

> *the attempt of certain quarters to explain their opposition to détente by hypocritical concern for Soviet Jews should be singled out especially. Under the false slogan of protecting Soviet Jews, these quarters are coming out against any agreement between the United States and the U.S.S.R. and in favor of an arms race. We Soviet Jews do not need to be protected from anyone or anything in our own country.*
>
> *... vicious slanders are being heaped about our country to distract the attention of the United States citizens from the struggle for peace and disarmament.*

The Soviet argument was thus twofold. First, American Jews were accused of using the Soviet human rights and Jewish emigration movements to incite anti-Soviet sentiment and prevent arms control. Second, it was argued that peace and arms control is of such paramount importance that reasonable people on both sides should not allow any other concerns, such as human rights, to get in the way. Soviet scientists, of course, understood from the very beginning the fallacies in the official Soviet position. They already understood and accepted the concept of linkage between science and democracy, much in the same way it was understood by the American revolutionaries of two centuries before. They then made further linkages to human rights and world peace. Thus, contrary to the Soviet government argument that world peace is the only issue that matters, it was the scientists who understood and argued that peace and human rights and democratization – are integrally linked, through the scientific paradigm, and must advance together.

Since so many of the democratic activists and refuseniks were scientists, the Soviet government's crackdowns on dissent had a dramatic effect on the conduct of the scientific enterprise in the Soviet Union. The scientist-activists, even when they managed to avoid being sent to prison or to a labor camp in

Siberia, were typically removed from their scientific positions, denied access to libraries and laboratories, and some even had their advanced degrees revoked. In many instances, in order to erase their very identities, citations to published work were removed from citation indices. Scientists had their papers and books confiscated in frequent raids at their homes by the Soviet secret police, the KGB. All of these measures were designed to isolate and discredit them and the concept of linkage that they espoused.

In the face of these obstacles to their scientific careers, the scientists tried to maintain scientific contacts by organizing clandestine scientific meetings and conferences in private apartments. The Moscow Science Seminar grew out of this effort. The Seminar met every Saturday morning at a different place in Moscow. These seminars were informal get-togethers at first, intended as a forum in which scientists could meet with other scientists and exchange scientific ideas, but as KGB harassment increased so did the organization of the Seminars and support from Western scientists. Many American and West European scientists during visits to the U.S.S.R. to attend "official" scientific meetings or exchange programs made a special effort also to attend the Moscow Science Seminar, even in the face of warnings from Soviet authorities not to do so. As support from Western scientists grew the Seminars were expanded into four International Conferences on Collective Phenomena, held in Moscow, with formal Proceedings papers published by the New York Academy of Sciences.[10]

The KGB eventually halted the Conferences altogether. Note that the role of the KGB in combating heresy and crushing dissent was not unlike that of the Inquisition in doing the same for the seventeenth-century Spanish Empire, as discussed in Chapter 1. The difference for the twentieth-century Soviet Union was the support provided for the Soviet dissidents by Western scientists. This support was significant. It led to the organization of a massive, worldwide scientist's boycott of the U.S.S.R., which, as I show in Chapters 7 and 8, was instrumental in bringing about the reforms in the Soviet Union that led to the ending of the Cold War. First, however, I discuss in the next chapter Sakharov's role in organizing and leading the protest movement that culminated in these reforms. Then in Chapters 5 and 6, I discuss the development of linkage in U.S. policy and the opposition to linkage of various groups in the U.S.

1. Here I specifically refer to the third Russian Revolution, the one of October-November 1917 that brought the Bolshevik (majority) Communist Party to power. The first Russian Revolution grew out of a protest movement by disparate groups of workers and students against the war with Japan in 1905. Russia was woefully unprepared militarily for the conflict with a growing Japanese power in the Far East, and gross mismanagement and misjudgment by the Russian Imperial government led to devastating military losses and dreadful economic deprivations at home. Nearly the entire Russian Navy was destroyed in a single afternoon's battle, for example. Despite these losses, the Russian government was able to suppress this first Revolution because the revolutionaries were not organized and were unable to win over the army or the general population. The Russian monarchy, furthermore, had some influential sympathizers in the West who saw a strong Russia as a necessary counterweight to the growing power of an imperial Germany in Europe. In fact, U.S. President Theodore Roosevelt was awarded the Nobel Prize for Peace for his role in negotiating an end to the Russo-Japanese War. For the revolutionaries, opportunity presented itself again, however, as a result of the far greater losses suffered by the Russian people in the First World War. Again, many of these disasters could be attributed to gross incompetence

on the part of the Romanov ruling family. The February Revolution (1917) swept away the monarchy with the promise to withdraw Russia from a European war in which the revolutionaries thought Russia had no real stake. When the new government failed to deliver on its promises to end Russia's involvement in the war it, in turn, was swept away when the Bolshevik (majority) faction of the Communist Party seized control of the government in the October Revolution. This third revolution was successful because of the organizational leadership of Vladimir Lenin, and, most significantly, by the winning over of the army to the side of the Communists. This latter accomplishment was largely the work of Leon Trotsky, who later became the author of one of the best, if not unbiased, eyewitness historical accounts of the Russian Revolution. When General Russky, commander of the Russian First Army in the First World War, came over to the Communist side, the success of the Communists in the Civil War that followed the October Revolution was assured.

The Western powers opposed the Revolution, and not just on ideological grounds. The Revolution had removed the eastern front in the allied war against the German and Austro-Hungarian Empires, and thus made it more difficult to bring this conflagration to a successful end. Indeed, the U.S. and its West European allies sent expeditionary forces to Russia to help the "Whites" in the civil war against the "Reds in an unsuccessful attempt to overturn the Revolution. The Cold War, however, did not really start until after the Second World War had left a divided Europe in its wake, with Soviet armies occupying essentially all of Europe east of Berlin and facing the armies of their former World War II allies. The Cold War was thus a military face-off that threatened to become a military clash, either in Europe or in various other places in the world where the interests of the United States and the Soviet Union collided. It was ideology backed up by military might that defined the parameters of the Cold War.

2. Vladimir Mayakovsky (1893-1930) was a well-known writer in pre-Revolutionary Russia, producing a large and diverse body of work, including poems, plays, and film scripts. He was a strong and ardent supporter of the Communist Party during the Revolution and Civil War, producing propaganda posters and literature for the Party. He gained his place in history with his eulogy for Lenin in 1923 with the now-famous lines "Lenin lived, Lenin lives, Lenin will live." His relationship with the Party was, however, complex and tumultuous. Some of his satirical works ran afoul of Soviet censorship, seemingly violating the new doctrine of "socialist realism". Mayakovsky became increasingly disillusioned with the direction of the new Soviet regime under Stalin, finally resolving for himself the intractable dilemmas he faced in his personal, political, and professional lives by committing suicide in 1930 at age 36.
3. See, for example, *Soviet Dissidents: Their Struggle for Human Rights*, Joshua Rubinstein (Beacon Press, Boston USA, 1980), which, in addition to being an excellent work on the subject, includes an extensive and excellent bibliography.
4. *The Washington Post*, Sunday, December 11, 1983.
5. Announced in *Pravda Ukrainy*, April 1, 1983. Pundits in the West could not help noticing that this was April Fool's Day.
6. See. for example, *The Antizionist Committee of the Soviet People* by Betsy Gidwitz, published by the Union of Councils for Soviet Jews, Washington, D.C., 1983.
7. The difficulty Jewish students had in gaining entry into the U.S.S.R.'s top school, Moscow State University, especially in Mathematics, has been discussed at length in various American professional society publications. Antisemitic propaganda in the U.S.S.R. has been well documented

by, for example, Dr. Ivan F. Martynov in the *samizdat* (underground) publication "The Korneyev Case: A Relapse Into the Propaganda of the Black Hundreds."

8. Soviet Jewish refusenik Dr. Iosef Begun, for example, was arrested for violating internal passport regulations as he returned home after his first term of Siberian exile.
9. *The New York Times*, Tuesday, June 7, 1983.
10. See, for example, "Fourth International Conference on Collective Phenomena," Annals of the New York Academy of Sciences Vol. 373, Edited by Joel L. Lebowitz, published by the New York Academy of Sciences, New York, New York, 1981.

Chapter 4

Sakharov

In 1963 Premier Nikita Khrushchev for the Soviet Union and President John F. Kennedy for the United States signed the Limited Nuclear Test Ban Treaty. This first nuclear arms control treaty banned the testing of nuclear weapons in the earth's atmosphere and outer space. Not only was this event a significant advance in arms control, but it was a spectacular personal victory for Dr. Andrei Sakharov. As a technical advisor to the Soviet delegation, he convinced the Soviet negotiators to revive an old American idea of treaty monitoring by "national technical means" that was ultimately the key to concluding the agreement. Sakharov was motivated to end atmospheric nuclear testing because he was very concerned about the worldwide health risks posed by radioactive fallout from these tests. He was also convinced that, with the latest Soviet developments in intercontinental ballistic missile technology, there was no longer any disadvantage in strategic parity for the Soviet Union in being restricted to underground testing.

"National technical means" of treaty monitoring meant that each side would be allowed to deploy earth-orbiting artificial satellites instrumented to detect clandestine nuclear explosions. Technical advances that had already been made by the early 1960s in remote imaging technology – in optical light and in x-rays and gamma rays – as well as in satellite launch technology, made such treaty monitoring possible. The idea was that technology made it no longer necessary to rely totally on blind trust in verifying treaty compliance. Sakharov effectively used arguments based on technology to advance the cause of arms control.

By 1963 Sakharov had already become a one-man constituency

for nuclear arms control. He had influence with the Soviet government when no one else in the country did solely because of his stature as the Soviet scientist who gave the U.S.S.R. its thermonuclear weapons capability.

He began his scientific career working under the direction of Igor Kurchatov at the new scientific institute founded just outside Moscow that was dedicated to exploiting the new discoveries in nuclear fission. Kurchatov had been given the job by Stalin to develop a nuclear weapons capability for the U.S.S.R., and was set to work under the watchful eye of Stalin's KGB chief, Lavrenti Beria. The work on bomb design moved later to the site of an old medieval monastery in the small Russian city of Sarov, where Sakharov spent the early years of his career in science. While Stalin was still alive the Soviet nuclear weapons development program could not risk any mistakes. Kurchatov and the other senior people knew precisely what fate was in store for them if a Soviet design were to fail on testing, so they chose the safe route of basically copying U.S. designs for the first fission bombs[1] that Beria obtained through espionage and forwarded to Kurchatov.[2] This is not to say, however, that Soviet scientists had no ideas of their own. Indeed, Sakharov was pursuing a radical new idea for a thermonuclear bomb, i.e., a bomb in which the fission "primary" serves as a trigger for the more powerful fusion "secondary." The U.S. nuclear weapons program was also working on ideas for thermonuclear (fission plus fusion) weapons, but the U.S. program was making little progress in getting their ideas for a "super" to work. Indeed, the basic U.S. idea for a "super" proved to be unworkable. Sakharov had a totally different idea that allowed the Soviet program to advance more rapidly to a practical design for a multi-megaton weapon, one that could be made small enough to be carried in an airplane or on a missile. The Soviet Union successfully tested its first thermonuclear weapon in 1955, just three years after the first U.S. test of a thermonuclear weapon.[3]

Thus, Sakharov did nothing less than provide the means for the Soviet Union to establish nuclear weapons parity with the U.S., and thus be considered an equal world superpower that had to be taken seriously in world affairs. For this achievement, he was considered a hero by his government. At the same time, though, he had begun to study and understand more and more about the effects of nuclear weapons explosions, and became more and more alarmed at what he saw as the extreme dangers of global nuclear war.

The "prompt" effects of atmospheric nuclear explosions were already well known. These include the effects of the prompt ionizing radiation (gamma rays, x-rays, and neutrons), the thermal (heat) wave, and the blast wave on people and structures in the vicinity of the explosion. What Sakharov came to worry about, however, were the longer-term effects of a massive exchange of nuclear weapons in a global nuclear war. For example, he realized that the thermal pulse would ignite continuous forest fires over wide areas that would put such a large quantity of smoke into the stratosphere over such a long period of time that the transparency of the earth's atmosphere to visible solar radiation would be reduced. The long-term effects on the earth's climate would be catastrophic, leading to a general global cooling. This idea of "nuclear winter" was further developed quantitatively by the U.S. scientist Carl Sagan and his colleagues in 1983.[4] Not only would the energy balance of the earth be changed, but the chemistry of the atmosphere would also change with the destruction of the stratospheric ozone layer. In addition, a global nuclear war would lead to disruption of transportation and communication systems, disruption of world food production and distribution networks, public utilities, medicine, and clothing. The consequences would be widespread famine and disease epidemics, universal chaos and social instability that could last many years, if not many generations. In short, a nuclear war between the superpowers would, in Sakharov's analysis, result in the total destruction of global civilization, and was, therefore, to be avoided.

Sakharov advanced the view that the balance of power between the U.S. and the U.S.S.R. was best maintained by working toward parity in conventional forces, rather than by using the threat of nuclear war to counter threats of conventional war aggression. He knew that U.S. strategic doctrine in defending Western Europe against an attack by Soviet and Warsaw Pact armies depended on nuclear weapons to counteract Soviet conventional force superiority. He was thus urging the Western countries to increase their conventional armaments in exchange for a decrease in nuclear weapons on both sides. Just as he had created the Soviet nuclear deterrent in the first place to help the Soviet Union achieve strategic parity with the West, he was now a strong advocate for the control of these weapons in order to secure world peace. Nuclear arms control could be achieved, he understood, through treaties in which violations could be effectively monitored.

Sakharov came to understand, however, that treaty monitoring by itself was not enough. Why is this so? The answer lies in the oft-misunderstood distinction between monitoring and verification in arms control. Monitoring is the process of making scientific measurements and collecting data. Verification involves analyzing these data, interpreting them, and then making a *political* judgment on how valid this interpretation is with respect to the treaty provisions. A case in point is the allegation that the Soviets used a toxic agent commonly called "yellow rain" to subdue rebellious villagers in Laos, Kampuchea (Cambodia) and Afghanistan in the early 1980s in violation of the 1972 International Convention on Biological and Toxin Weapons to which they were a signatory. The U.S. Senate, in fact, passed a resolution unanimously in February 1984 condemning Soviet use of yellow rain toxins in Laos and Afghanistan. The Soviets consistently and repeatedly denied the charges. In the alleged yellow rain incidents, much of the evidence consisted of eyewitness accounts. Even though many of these accounts were remarkably parallel and consistent with the interpretation of yellow rain use, eyewitness accounts are notoriously unreliable and unscientific, and as such cannot be used to constitute proof of anything. As a result, the U.S. government went to extraordinary lengths to obtain a physical sample of the stuff to be analyzed in the laboratory. Even after this was done many U.S. scientists were critical of such definitive conclusions drawn from one tiny sample, removed from its natural environment and tested in the absence of a control.[5] Others expressed skepticism that the Soviets could or would employ such toxins in remote mountain villages.[6] Still, others felt that one would have to be willfully obtuse to ignore the overwhelming preponderance of evidence, even though circumstantial, that the Soviets had violated the chemical and biological weapons treaty. This difference of opinion points up a fundamental dilemma for arms control negotiators. This dilemma is that no arms control treaty can be truly verifiable even though it can be monitored. Monitoring is useful in arms control, however, when there is free and open communication on both sides, that is, when citizens are free to monitor their government's compliance and to hold it accountable. Sakharov understood this fundamental truth by the early 1960s. Thus, he concluded that arms control was possible only with democratization of the Soviet Union.

The linkage of arms control with democratization was a crucial insight. It was then a relatively easy step to make the further linkage with human rights. It was painfully obvious to Sakharov that there were all sorts and sizes of citizen-constituencies for arms control in the Western countries that had no counterparts at all in the Soviet Union because Soviet citizens were not free to express their opinions, come together in groups and associations, publish, meet with foreigners, or lobby their government, let alone criticize it. Without these fundamental freedoms, which are exactly the same freedoms required for science to flourish, arms control and world peace, according to Sakharov's thinking, were impossible to achieve. "The opportunity to criticize the policy of one's national leaders in matters of war and peace as you do freely is, in our country, entirely absent," wrote Sakharov in 1983 to the American scientist Sidney Drell.[7] He concluded the letter to Drell with the following thoughts.

> *I again stress how important it is that the world realize the absolute inadmissibility of nuclear war, the collective suicide of mankind. It is impossible to win a nuclear war. What is necessary is to strive, systematically though carefully, for complete nuclear disarmament based on strategic parity in conventional weapons. As long as there are nuclear weapons in the world, there must be strategic parity of nuclear forces so that neither side will venture to embark on a limited or regional nuclear war. Genuine security is possible only when based on a stabilization of international relations, a repudiation of expansionist policies, the strengthening of international trust, openness and pluralization in the socialist societies, the observance of human rights throughout the world, the rapprochement – convergence – of the socialist and capitalist systems, and worldwide coordinated efforts to solve global problems.*

Sakharov wrote these lines during the course of his internal exile in the closed industrial city of Gorky, where he was sent in January 1980 and kept under house arrest and close surveillance for having publicly criticized the 1979 Soviet invasion of Afghanistan. The invasion was a shock to Sakharov and the dissident community because it clearly signaled that the forces of liberalization and democratization in the Soviet Union were losing. If anything, repression was intensifying, and relations with the West were reaching a new low point. Sakharov clearly had the Afghanistan invasion in mind when he included "repudiation of expansionist policies" in his list of reforms necessary for nuclear disarmament and world peace. By the time his exile began, however, he was already nearly 59 years old, and over the course of his exile and isolation he would suffer irreparable deterioration of his health that impaired his ability to keep up the fight for these reforms.

This attack by the regime was therefore more serious and more damaging than the earlier one. As already mentioned, the ideas expressed in Sakharov's letter to Drell were already taking shape in the 1950s and 1960s, maybe even earlier.

Sakharov was born into an educated family on 21 May 1921. His father was a physics teacher, and the author of some popular physics books. After graduating from high school in 1938 Sakharov enrolled in the Physics Department at Moscow State University, and graduated with honors in 1942. By this time the Second World War was well underway, and university classes had been moved inland to Ashkhabad. After graduation, he was first sent to a remote area to assist with the war effort by working on a logging operation. This first encounter and experience with ordinary workers and peasants in the countryside had a profound effect on him. It was perhaps here that the first seeds were planted of his later human rights advocacy. It was also during this time that he met Klavdia Vikhireva in Ulyanovsk. They were married on July 10, 1943. He and Klavdia had three children together.

After the war ended, he went for graduate studies to the Lebedev Institute of Physics in Moscow, where he studied under the direction of one of the Soviet Union's most prominent theoretical physicists, Igor Tamm. He and Tamm were recruited to work on the Soviet bomb project in 1948. "We were all convinced of the vital importance of our work for establishing a worldwide military equilibrium," wrote Sakharov in 1981,[8] "and we were attracted by its scope." Sakharov and his family "disappeared" into the secret world of Soviet nuclear weapons work, which had been relocated to a laboratory site in the medieval monastery town of Sarov.

During his years in the Soviet nuclear weapons program, his thoughts on nuclear weapons, science, politics, and peace underwent a transformation. At least two events during this time helped to propel this transformation. One was the first test of his thermonuclear weapon design, in 1955. According to Sakharov's own account,[9] there were human casualties resulting from this test, which, among other things, resulted in a confrontation between Sakharov and Soviet Marshal Nedelny. The more lasting impact came from his confrontation with Khrushchev when, in 1961, Sakharov criticized the Soviet leader for violating the nuclear test moratorium that had been in effect since 1958.

He first published his thoughts in early 1968 in a *samizdat* essay entitled "Progress, Coexistence, and Intellectual Freedom." Copies of this publication were widely distributed in the Moscow intellectual community, and copies found their way to the West where the essay was reprinted and widely read. By publishing his essay Sakharov wished to stimulate an open discussion of his ideas in the hopeful atmosphere of the "Prague Spring", when reformists who came to power in Communist Czechoslovakia raised hopes in Moscow that reform would be possible there, too. Such hopes were decisively crushed when Soviet Army tanks rolled into Prague in August of that year. Of course, the Soviet government could not destroy dissent in Czechoslovakia without doing so also in Moscow. Sakharov, who had received three Orders of Socialist Labor – the highest civilian honor in the U.S.S.R. – the Stalin Prize and the Lenin Prize for his work in nuclear weapons design, now lost his security clearance and his job, along with all the perquisites that he had enjoyed up until then. These included special housing, a chauffeur, access to consumer goods unavailable to the ordinary Soviet citizen, and even a bodyguard.

Instead of the open debate and discussion that he had hoped his essay would provoke, what he received instead from the state was rejection and isolation. This was just two years after he signed his name to the petition to Brezhnev, as mentioned in the last chapter, his first public expression of dissent. A further blow came the following year when his wife died on March 8 of stomach cancer.

Then, just two years after the publication of his essay he and two other democrats founded the Moscow Committee for Human Rights. People who had formed organizations prior to this time were met with immediate, and harsh, reprisals, so the founders of the Moscow Committee for Human Rights surely knew of the dangers for themselves in taking this step. The dilemma was that they could not hope to advance the cause of human rights without coming together in association, but if they did come together then the reaction of the authorities would prevent the very progress they hoped to achieve. This time, however, Sakharov and his colleagues created the dilemma for the state, because instead of a direct provocation and challenge to state authority, the Committee was chartered as "a creative association acting in accordance with the laws of the land," offering "creative assistance to persons engaged in constructive research into the theoretical aspects of the human rights question and in the study of the specific nature of this question in a socialist society." In other words, Sakharov was making clear that this new organization was not really an organization at all, and certainly not one composed of radical activists determined to damage or overthrow the Soviet system. Instead, this association was meant to be a study group, one that would act as a kind of legal aid society addressing theoretical issues only. Within this framework, it was then possible to attract the support and attention of all sorts of people. The main accomplishment of this approach to the issue of democratic reform was in providing the theoretical underpinnings for all the disparate elements of democratic reform in the Soviet Union. Sakharov and his colleagues on the Committee spent their time advising people on human rights issues, attempting to attend closed political trials, and informing Western journalists about the specifics of particular cases.

After the Helsinki Accords were signed in 1975 Sakharov became a founding member of the Public Group to Promote Observance of the Helsinki Accords in the U.S.S.R. The Group's first press conference was held in the Moscow apartment that Sakharov shared with his second wife, Yelena Bonner, a divorcee with two children who he had married on January 7, 1972. There, he introduced to the press fellow physicist Yuri Orlov as the leader of the new Group. This Helsinki Watch Group further accelerated the process started by the Moscow Committee for Human Rights in bringing together in common cause the democratic reformers, the leaders of the Jewish emigration movement, and activists for repatriation of Crimean Tatars[10] and other human rights activists. The leaders of the new combined effort were all scientists – physicists Sakharov and Orlov, and a young computer scientist named Anatoly Shcharansky. This was no accident. It was the scientists who provided the intellectual glue that held it all together.

For all these accomplishments Andrei Sakharov was awarded the Nobel Prize for Peace in 1975. As with Pasternak before him, the Soviet government pressured Sakharov not to accept the award, considering it an insult to the Soviet Union, and would not issue him a visa to leave the country to accept the award. As it turned out, Sakharov's wife Yelena Bonner was in Italy undergoing medical treatment in late 1975. An American woman visiting Moscow on a tourist visa met clandestinely with Sakharov and Shcharansky; she then left the country with Sakharov's Nobel Prize speech hidden in her bra,[11] delivered the manuscript to Mrs. Bonner in Italy, who then flew to Oslo to accept the Prize on behalf of her husband and deliver his speech for him.

The Soviet government enlisted other Soviet scientists in their campaign to discredit Sakharov and his ideas. Since all Soviet scientists depended on the state for their jobs, coercion to enlist their cooperation was an easy matter. Accordingly, 72 members of the Academy of Sciences of the U.S.S.R. signed a letter published in the official Soviet government newspaper *Izvestia* (News) on October 26, 1975, to "express our bewilderment and indignation in connection with the decision of the Nobel Committee." The full text of this letter appears in Appendix I along with the names of the 72 signatories. Among the signatories are Soviet winners of the Nobel Prize in Physics, like Nikolai Basov. About two years earlier a smaller number of Academicians published a more general attack on Sakharov in the Communist Party newspaper *Pravda* (August 29, 1973). Since it is important to understand the Soviet government position with respect to Sakharov – and how it was mirrored by some people in the West – we will examine the text of this letter.

> *We consider it essential to bring to the attention of the general public our attitude regarding the activities of Academician A.D. Sakharov.*
>
> *In the last few years, Academician Sakharov has moved away from active scientific activities and spoken out with a series of pronouncements against the Soviet government's internal and foreign policy. Not long ago in an interview given by him to the foreign correspondents in Moscow and published in the Western press, he went so far as to speak out against the détente policy of the Soviet Union and against the policy of consolidating those positive steps which have taken place in the whole world recently.*
>
> *These announcements of A. D. Sakharov, deeply alien to the interests of all progressive people, attempt to justify a crude*

distortion of Soviet reality and imaginary criticisms of the socialist system.

In his statements, in essence, he allies himself with the most reactionary imperialist circles, actively speaking out against efforts to bring international cooperation among countries with different social systems; against the policy of our party and our government supporting the development of scientific and cultural cooperation; and the consolidation of peace among peoples. In the same way Sakharov has actually become the instrument for hostile propaganda against the Soviet Union and other socialist countries.

The activity of A. D. Sakharov is fundamentally alien to Soviet scientists. It looks particularly unseemly in light of the concentration of our efforts on solving the vast problem of economic and cultural structure of the U.S.S.R., on strengthening peace and improving the international situation.

We want to express our indignation with the pronouncements of Academician Sakharov, and we emphatically condemn his activity, which discredits the honor and the dignity of the Soviet scientist.

We hope that Academician Sakharov will meditate on his activities.

Not only are these statements clearly distortions and mischaracterizations of Sakharov's views, but as "an instrument of hostile propaganda" he is accused of being an enemy of socialism, of the U.S.S.R., and of world peace. These statements, of course, represented the official Soviet view, so it can be claimed that the signatures of the Academicians on the statements were not given voluntarily. Since their careers and positions depended entirely on the support and favor of the state (as did Sakharov's), it is virtually certain that coercion of one sort or another, even if subtle and unspoken, was employed to obtain the signatures on these statements. This is not to say, however, that the signers did not agree with the sentiments expressed therein. It is impossible for us to know exactly what motivated each person to consent to having his name on these statements at that particular time. Not one of the signers has since publicly recanted his anti-Sakharov views, even though now nearly every Russian scientist will not let an opportunity pass to express to a Western colleague his respect and admiration for Sakharov. Whether these more recent statements, offered privately, are any more sincere than the earlier public ones it is also impossible to say.

It is worth noting, though, that most Soviet scientists and Academicians did not sign anti-Sakharov statements. Indeed, a move to oust Sakharov from membership in the Soviet Academy of Sciences failed to get a majority vote of Academy members. One of the instigators of the oust-Sakharov effort was Nikolai Basov, who shared the Nobel Prize in Physics in 1964 (with his colleague Alexander M. Prokhorov – also a signatory to both the anti-Sakharov letters published in the Soviet press – and the American scientist Charles H. Townes) for the invention of the laser. His name appears on both anti-Sakharov statements. Basov was the Director of the Physical Institute of the Academy of Sciences in Moscow (commonly referred to as the Lebedev Institute), the same Institute where Sakharov was employed before his exile to Gorky. After Mikhail Gorbachev became Communist Party Chairman, one of the *perestroika* reforms he instituted was direct democratic election of Institute and Laboratory directors by the scientific staff (rather than selection by the Party), and in the first such election Nikolai Basov was himself ousted as Director of the Lebedev Institute.

But we are getting ahead of the story. The point here is that most Soviet scientists probably did share Sakharov's basic views, particularly his views on scientific freedom and its dependence on free and open communications and contacts with the wider world of science. They also largely had the same understanding that Sakharov did about the linkage of these freedoms to arms control and peace. It is also important to note that Soviet scientists played the leading roles in all the human rights, peace, and democratization movements in the U.S.S.R., as well as in the Jewish emigration movement. Even those who were not so involved were in favor of maintaining contacts with foreign scientists, and did not consider themselves "instruments of hostile propaganda" for holding such views. They did, however, place themselves at some risk in publicly expressing such views. Most who did were treated quite harshly, having to serve time in prison, labor camp, or (like Sakharov) in internal exile.

It was not enough, however, for the regime to silence its outspoken scientists. The Soviet government also felt compelled to discredit them and their views. This was the point, of course, of the Academicians' statements and the work of the Anti-Zionist Committee.

The Soviet government, of course, was right to feel threatened. Since Communist Party rule came to depend for its very existence on the imposition of police state controls, the scientists' agitation for the loosening of such controls did indeed threaten the very existence of the state. Thus, it was critically important to the state's survival to discredit the scientist's ideas about linkage. Soviet propaganda was largely directed to this end. The intended audience for this propaganda was not principally the Soviet public. Since there were no independent political associations or constituencies in the Soviet Union that could lobby the government – the Soviet scientific community was the closest approximation to such a lobby for arms control and democratization – the Soviet public did not need to be won over. They could be controlled. The American public, and particularly those segments of the American public that could exert influence on the policies of the U.S. government, did need to be won over. It was principally the American public that was the intended target of Soviet anti-linkage propaganda.

Soviet propaganda campaigns, in any case, were unneeded

effort. Large segments of the American public, on both ends of the political spectrum, were already opposed to linkage for their own separate reasons. Indeed, for many Americans, there was no contradiction between being repelled by Soviet anti-linkage propaganda and being opposed themselves to linkage. It was ultimately the scientists' efforts in promoting this idea that were decisive – not in winning over the general U.S. public to the idea – but in winning over the Soviet government. The Cold War effectively ended when Mikhail Gorbachev freed Andrei Sakharov from internal exile and accepted his counsel on the linkage of democratic reform, economic modernization, and world peace.

Once again, though, we are getting ahead of the story. Before we discuss the pressure brought to bear on the Soviet government by the scientists' boycott, and Soviet responses to it, we first examine opposition to linkage in the U.S.

1. Fission is the process by which a heavy nucleus, such as uranium, splits, or fissions, into two approximately equal-mass halves, with an energy release that is equal to the difference in the rest mass energies of the parent nucleus and the two "daughter" nuclei. Most of the released energy is carried away by the kinetic energy (energy of motion) of the daughter nuclei. Fusion, on the other hand, is the process by which two light nuclei (usually isotopes of hydrogen, the lightest element) are brought together under conditions of very high pressure and temperature to fuse together into a heavier nucleus with the release of an energetic subatomic particle which carries away most of the energy. In the fusion of deuterium and tritium (two heavy isotopes of hydrogen), for example, most of the energy produced in the fusion reaction is carried away by energetic neutrons. The interested reader can find a basic unclassified description of the physics of nuclear explosions in *The Effects of Nuclear*

Weapons by Samuel Glasstone, originally published by the U.S. Atomic Energy Commission in 1962 and available from the U.S. Government Printing Office, Washington, D. C. 20402.

2. Soviet espionage during the early days of the U.S nuclear weapons program, and its importance to the development of the Soviet nuclear weapons program, has been discussed extensively in numerous books and articles. A particularly comprehensive account can be found in Dark Sun: *The Making of the Hydrogen Bomb* by Richard Rhodes, Simon and Schuster, 1995.
3. I am counting here the Mike test of the Ivy series of nuclear tests in the Pacific as the first U.S. test of a thermonuclear weapon design, conducted on October 31, 1952. The Mike device used liquid deuterium as its fusion fuel, and thus incorporated complex and bulky cryogenic hardware as part of the test device. Since deuterium is a gas at room temperature, the cryogenic hardware was required to liquify it. The Mike test had a yield equivalent to that of 10.4 megatons of TNT. The U.S. tested a design using solid lithium deuteride as a fusion fuel in 1954, a few months before the Soviet test of Sakharov's design, but it was Sakharov's design that pointed the way to a more compact and workable design that could be carried on intercontinental ballistic missiles.
4. "Nuclear War and Climate Catastrophe," by Carl Sagan, *Foreign Affairs* Vol. 62/2, Winter 1983/1984.
5. An excellent account of the entire "yellow rain" controversy is provided by Robert L. Bartley and William P. Kucewicz in " 'Yellow Rain' and the Future of Arms Agreements," *Foreign Affairs* Vol. 61/4, Spring 1983.
6. Saul Hormats, *The Washington Post National Weekly Edition*, March 12, 1984.
7. Taken from a letter by Andrei Sakharov published in *Facing*

the Threat of Nuclear Weapons by Sidney D. Drell, Univ. of Washington Press, Seattle, 1983.

8. "The Social Responsibility of Scientists," by Andrei Sakharov, *Physics Today*, pages 25-30, June 1981.
9. This account is contained in an open letter that Sakharov wrote to the President of the Soviet Academy of Sciences, A. P. Aleksandrov, dated October 20, 1980, from his exile in Gorky, and translated by Richard Lourie.
10. Crimean Tatars are a Turkic-speaking ethnic group indigenous to the Crimean Peninsula since probably the tenth century. They were converted to Islam during the rule of the Mongol Golden Horde in the 14th century. Crimea became one of the centers of Islamic civilization in Eastern Europe, with the Tatars comprising the majority of Crimea's population into the middle of the 19th century. Immediately after the Soviet Army recaptured Crimea from the Axis powers in May 1944, Stalin considered that all Tatars were Nazi collaborators (most were not), and had the whole lot of them deported to Uzbekistan in boxcars.
11. The author learned about the smuggling of Sakharov's Nobel Prize acceptance speech directly from the woman who did the smuggling, but she never consented to being publicly identified, so her privacy will remain protected.

Chapter 5

Resistance to Linkage on the American Right

The American right has always seen arms control as largely inimical to U.S. interests. Instead of seeking parity in conventional armaments, as Sakharov advocated, and managing the Cold War conflict through arms control agreements, the right sought military dominance. In fact, their view has been that world peace could only be guaranteed through U.S. military strength. Arms control advocates, many of whom were scientists, were accused of ignoring the very real threat posed to the Western democracies by a strong Soviet military power. Thus, seen in this light, arms control was often viewed from the right as, at best, appeasement of the enemy and, at worst, surrender.

In the early years of the Reagan administration, no one appeared to be a stronger and more eloquent spokesman for this point of view than President Ronald Reagan himself. He was elected to the Presidency in 1980 partly on the promise that he would reverse the decline in spending on the U.S. military that he claimed had taken place under his Democratic predecessor, President Jimmy Carter. In fact, though, the U.S. military buildup of the 1980s actually was put in motion by President Carter in 1979 in the wake of the Soviet invasion of Afghanistan. Even before the Soviet invasion of Afghanistan defense spending rose every year during the Carter Presidency.[1] Nonetheless, it is true that President Reagan, with a Republican Senate supporting him for six of his eight years in office, did escalate the U.S. military buildup. The President backed NATO's installation of intermediate-range missiles carrying nuclear warheads in Europe as a counter to the Soviet's deployment of their new SS-20 missiles aimed at Western Europe. His 1984 proposal of a

Strategic Defense Initiative (SDI) – development and deployment of a defensive shield to protect the U.S. against nuclear missile attack – was seen in arms control circles as highly destabilizing and a dangerous escalation of the superpower arms race.

The right, however, did not see SDI the same way. To them, it seemed clear that the U.S. would always prevail in the end because of an assumed American scientific and technological superiority, and because the U.S. could simply outspend the Soviets. Along with the assumption that the U.S. *could* prevail came the moral imperative that the U.S. *should* prevail. Thus, for the right ending, the Cold War was not the end goal; winning it was. They took great offense, however, in being characterized as warmongers. They were truly committed to the idea that Communism was an unmitigated evil, and only victory over Communism could assure a stable peace.

There are many on the right who claim today that Reagan's peace-through-strength confrontationalism is what won the Cold War by forcing the Soviet Union to spend itself into bankruptcy. These claims, however, do not withstand scrutiny.

Any serious examination of Soviet responses to Reagan's belligerent rhetoric and the U.S. military buildup shows that while there was concern and worry among Soviet military planners there was no panic, no over-reaction. There was certainly no commensurate Soviet military buildup. During the first half of the 1980s Soviet policy was largely directed at maintaining the status quo. Besides, the Soviet military was bogged down in what was, as was becoming increasingly clear to them, an unwinnable war in Afghanistan. The Soviets reacted to SDI more with confusion than with alarm. Even while putting forth their view that SDI would be a destabilizing threat to deterrence-based peace (because it would effectively give the U.S. a "first strike" capability by eliminating the threat of retaliation on which deterrence was based) they also believed that the American proposals were technically unworkable, and in any event, could be easily counteracted at far less expense than what the U.S. was proposing to spend to develop and deploy a defensive system. The fact that the Soviets did not spend even modest sums on SDI countermeasures, or on an SDI of their own, shows that the American right's claim that SDI forced the collapse of the Soviet Union is just nonsense.

Further, even by the time Mikhail Gorbachev came to power in the Soviet Union in March 1985 President Reagan was already softening his rhetoric, and was even proposing some radical arms control initiatives of his own. In Gorbachev, he even found a kindred spirit who shared his interest in the ultimate elimination of nuclear weapons, one of the primary goals of the American left. Cynics suggest that Reagan only made this move in order to preempt the growing Nuclear Freeze movement in the U.S.[2] It is more likely, as Frances Fitzgerald[3] has explained, that Reagan truly was interested in eliminating the threat of nuclear war, and that this interest came from his keen sensitivity to what the American people wanted. What the American people wanted, it seemed, was an end to the threat of nuclear war. Whether Reagan's pursuit of arms control with Gorbachev was sincerely or cynically motivated, the mere fact of his engaging the Soviets in arms control negotiations contradicts later claims by the right that the U.S. military buildup is what caused the collapse of the Soviet Union and the end of the Cold War.

Although the right professed to advocate human rights in the Soviet Union, they clearly neither understood nor applied the concept of linkage. Since linkage recognized the necessity of using advances in one area to leverage advances in the other, as Sakharov had proposed, and since they were fundamentally opposed to arms control, human rights advocacy was used by the right only as a cudgel with which to beat the Soviets and win points in the on-going East-West ideological debate of the Cold War. Human rights groups could not take seriously the criticisms the right leveled at the Soviet Union for human rights violations while the conservative governments in the U.S. and Western Europe continued to support and aid friendly third-world dictators responsible for even more dreadful human rights violations. Clearly, human rights advocacy was pursued only when it was consistent with the right's foreign policy goals, the primary one of which was the overthrow of Communism. Linkage, therefore, was opposed because linkage implied giving equal weight in formulating policy to all the linked goals – arms control, peace, democratization, human rights – whereas military superiority over Communism, and its overthrow, were, to the right, the prime goals to which all others were to be subordinated.

While ideological battles over arms control continued all through the Cold War, the principal opposition to linkage came from the U.S. business community. For American businessmen interested, for example, in exporting their commodities to a ready Soviet market, opposition to linkage was not ideological – as it was for arms control – but rather motivated by a practical consideration of commercial interests. It was in the arena of trade that the political battle over linkage was joined in the U.S. Congress. It was in the arena of trade also that the most visible evidence could be found of an official U.S. government policy of linkage. This linkage came with the enactment of Title IV of the Trade Act of 1974. This section of U.S. trade law, originally offered as an amendment to the Trade Act by Senator Henry M. "Scoop" Jackson (a Democrat from Washington State, first elected to Congress as a member of the House in 1941) and Representative Charles A. Vanik (a Democrat from Ohio, first elected to the House in 1955) and hence commonly referred to as the Jackson-Vanik Amendment, specified that no nation could receive Most Favored Nation (MFN) trade status, and the tariff benefits this entails, without the President certifying to Congress that its government allows emigration without restrictions and harassment. Specifically, paragraph (a) of Section 2432 of Title

IV declared that:

> *– products from any nonmarket economy country shall not be eligible to receive nondiscriminatory treatment (MFN treatment), such country shall not participate in any program of the Government of the United States which extends credits or credit guarantees or investment guarantees, directly, or indirectly, and the President of the United States shall not conclude any commercial agreement with any such country, during the period beginning with the date on which the President determines that such country –*
> *(1) denies its citizens the right or opportunity to emigrate;*
> *(2) imposes more than a nominal tax on emigration or on the visas or other documents required for emigration, for any purpose or cause whatsoever; or*
> *(3) imposes more than a nominal tax, levy, fine, fee, or other charge on any citizen as a consequence of the desire of such citizen to emigrate to the country of his choice.*

The remainder of the Amendment details the procedures for the President to authorize waivers of the application of these MFN restrictions for specific countries, and the responsibilities of Congress to continue or cease such authority.

The Jackson-Vanik Amendment was specifically directed at the U.S.S.R. and other East-bloc nations, particularly Romania, where restrictions on Jewish emigration were most severe. The intent of the Amendment was to use the promise of MFN trade status to leverage a loosening of emigration restrictions since the President was authorized to issue a waiver by Executive Order every twelve months. President Gerald Ford used the authority granted him under this Act to order a waiver for Romania in April 1975 (Executive Order No. 11854). President Jimmy Carter ordered waivers for Hungary in April 1978 (Executive Order No. 12051) and the People's Republic of China in October 1979 (Executive Order No. 12167).

The Amendment had the strong support of the Soviet dissident and Jewish communities in the U.S.S.R. as well as their growing number of supporters in many Jewish and independent Soviet Jewry organizations in the U.S. and various human rights groups. All these groups, after the Amendment became U.S. law on January 3, 1975, focused their efforts on assuring that the President did not order a waiver for the U.S.S.R. without "substantial evidence" of increased emigration. Since the President could order such a waiver every twelve months, the lobbying and public relations battle over linkage of trade and human rights was re-fought yearly.

The principal proponents of waivers were, of course, American businessmen, particularly farmers and agribusiness concerns, and their political supporters in Congress. Many Senators and Representatives from mid-west farm states were the leaders of the pro-waiver forces in Congress. They and other critics of linkage argued that the Jackson-Vanik Amendment was counter-productive because Jewish emigration from the U.S.S.R. actually declined after its passage into law. The critics' view was that trade, rather than a cutoff of trade, was a more effective lever to induce changes in Soviet behavior. Constructive engagement via trade, so the argument went, would expose the totalitarian state to "the American Way," and thus open channels for Americans to influence change in the Soviet Union.

The argument that the Jackson-Vanik Amendment was counter-productive is difficult to assess with any certainty because there were so many other factors influencing Soviet behavior with respect to Jewish emigration. It is true that after Jewish emigration from the U.S.S.R. began in 1969 the total number of people emigrating per year reached a high of about 35,000 in 1973, and then experienced a sharp decline beginning in late 1973 just as debate on the trade legislation was intensifying. It is more likely, though, that the sharp decline in emigration starting in 1974 was a direct response to the Yom Kippur War in the Middle East in October 1973 and the Arab oil embargo that followed. The Soviets, who tacitly supported the embargo, were pressured by their Arab client states to quit supplying additional manpower for the Israeli army. Israel was the only allowed destination for emigrating Soviet Jews, but since the U.S.S.R. and Israel did not have diplomatic relations there were no direct flights between the two countries. Emigrating Soviet Jews went first to Vienna or Rome and were re-routed from there by various Jewish aid groups. Over the years an increasing fraction of émigrés opted to go to the U.S. or Canada as their final destination, even though they carried Israeli visas. In 1974, the year the Jackson-Vanik Amendment was enacted, the fraction of émigrés opting out of going to Israel was nearly 19%, up from only 4.2% the year before. The total number of émigrés did not increase again to exceed the 1973 high until 1979, when more than 51000 Soviet Jews emigrated. In that year the fraction opting out of going to Israel was 65%. A chart showing the number of émigrés per year from 1969 to 1989 is contained in Appendix II.

Although 50,000 per year may appear to be a large number, it still represented only a small fraction of those who either had applied to leave and had their emigration applications rejected or refused, or those who wished to leave but did not or could not apply because of various arbitrary restrictions. The Union of Councils for Soviet Jews (UCSJ), the largest independent advocacy group in the U.S. that worked on behalf of Soviet Jews, had estimated that as much as one-half of the total Jewish population of the U.S.S.R. expressed a desire to leave the country. In the 1970s the U.S.S.R. had the world's second-largest Jewish population (after the U.S.), some 4 million people; to get out all those who wanted out would take sustained levels of emigration of 100,000 to 200,000 people per year over 10 to 20 years. The actual levels were nowhere near this high, and this is the principal reason why UCSJ and other human rights groups fought so hard against waivers of Jackson-Vanik. Their policy was that to earn a waiver the U.S.S.R. needed to show evidence of "substantial performance" on improving Jewish emigration. According to a UCSJ policy paper, "what the UCSJ believes would be evidence of 'substantial performance" is the expeditious commencement of a process that allows a continued and substantial number of persons to emigrate each year without the type of institutional interference that has pervaded and characterized the Soviet Union's actions and attitude toward emigration." They also called for the release of all Jewish and non-Jewish prisoners held for their human rights activities, and immediate emigration of all long-term refuseniks. Since none of these things came about in the immediate aftermath of Jackson-Vanik, critics of the Amendment argued that trade restrictions were counter-productive or, at best, ineffective. Proponents argued that, on the contrary, the restrictions could only be effective with sustained application.

Scientists were largely outside this particular debate over trade, but the arguments on both sides of the trade/no-trade debate are important because they would be raised all over again in the later debate over the scientists' boycott of the U.S.S.R. In all these debates the argument was not over goals but over efficacy. In the arms control debates, on the other hand, in which scientists were integrally involved – and not only because of their technical expertise – the arguments were very much over goals, and hence were much more complicated. Here, as we have already seen, the argument was between those who were ideologically opposed to arms control and linkage, and those who saw linkage as a prerequisite and necessary condition to arms control. Most scientists – but not all – were on the side of linkage in these debates. Scientists, individually and through their professional societies, worked to influence the U.S. government to keep to a steady course of pursuing a policy of linkage in its relations with the U.S.S.R. The scientists, though, did not only have to contend with opposition from the right. There was also plenty of opposition from the left.

1. The U.S. defense budget for FY 1981, the last budget passed during the Presidency of Jimmy Carter, was $ 159.7B appropriated (compared to the President's request of $154.5B), an increase of 22% over the previous year's defense budget. President Carter requested $200B for FY 1982 just before leaving office in January 1981, nearly a doubling of the defense budget since he took office in FY 1977. This despite the fact that he had campaigned in 1976 on a promise to cut the defense budget by $5B to $7B, and to withdraw the U.S. military from South Korea. Neither promise was kept. This information comes from "Congress and the Nation Vol. V 1977-1980," Congressional Quarterly, Inc., Washington, D.C.
2. *The New York Times*, May 2, 1982.
3. *Way Out There in the Blue* by Frances Fitzgerald (Simon and Schuster, 2000).

Chapter 6

Resistance to Linkage on the American Left

The Jackson-Vanik Amendment was perhaps the most visible evidence of a U.S. policy of linkage. In the area of arms control linkage was also the official policy of the U.S. through both Democratic and Republican administrations. In sharp contrast to the situation in the U.S.S.R. where the independent Helsinki Watch Groups were brutally suppressed, various non-governmental organizations and citizen groups in various cities around the U.S. which were strong advocates of linkage[1] and were working to monitor compliance with the Helsinki Accords had wide bipartisan support in Congress. In addition, a bipartisan Congressional Commission (established by Public Law No. 94-304, 90 Stat 661 on June 3, 1976) provided an official monitoring of all signatories' compliance with all provisions of the Accords. In this capacity, they were just as quick to point out American violations of the Accords as Soviet violations. When, for example, the U.S. State Department refused entry visas to several Soviet scientists to attend an international scientific conference in California in 1980, the American Physical Society and the Committee of Concerned Scientists protested the action to the State Department, citing the action as a violation of the Helsinki Accords.[2] The Commission agreed. Thus, the Commission established itself as a credible and honest treaty monitor, and its well-documented judgments[3] on the Soviet's frequent violations of all three Baskets of the Accords, including that dealing with confidence-building measures, could not be construed merely as U.S. government propaganda.

Besides some segments of the U.S. arms control, diplomatic, and business communities, the principal opposition to linkage in the U.S. came from the U.S. peace movement. The original objective of peace movement activists was unilateral disarmament; nuclear weapons became the focus of their efforts to begin achieving this objective. Many activists in the peace movement still cling to this objective. Many others abandoned it because it never received wide popular support. Instead, they embraced the concept of bilateralism. In this concept compliance to arms control agreements is assured because both sides are assumed to have an equal interest in the agreements' success. Besides, said the bilateralists, we do not have to put our trust in the Russians because we can always negotiate a treaty that is verifiable. The implicit assumption in this argument is that the Soviets would not dare violate a verifiable treaty. Thus, the bilateralists were opposed to linkage because proponents of linkage who were continually calling attention to Soviet human rights abuses, and therefore, verifiable violations of treaty agreements by the U.S.S.R., were destroying the validity of the basic assumption on which the bilateralists' whole argument depended.

Even less did the bilateralists wish to hear about Soviet violations of existing arms control agreements. We already discussed in Chapter 4 the specific case concerning allegations that the U.S.S.R. used a toxic agent commonly called "yellow rain" to subdue rebellious villagers in Laos, Cambodia, and Afghanistan in violation of the 1972 International Convention on Biological and Toxin Weapons to which they are a signatory. In the alleged "yellow rain" incidents all the difficult questions concerning treaty monitoring and verification were apparent. For example, what constitutes verifiability? What sort of evidence, and in what amounts, is necessary to prove non-compliance? Many people, particularly scientists, began to understand the truth of Sakharov's thesis that these questions could only be addressed adequately when there is free and open communication on both sides, when citizens are free to monitor their government's compliance and – most importantly – to hold it accountable. In the absence of such freedoms – symmetric on both sides – questions of treaty verification would always remain unresolved dilemmas.

These dilemmas were evident also in the debate over a bilateral nuclear weapons freeze. The American Peace movement embraced the nuclear freeze idea in the 1980s as their principal organizing idea, and worked very hard to turn their Nuclear Freeze campaign into a mass citizen movement. Proponents of a freeze on the testing, development, production, and deployment of nuclear weapons claimed that a freeze would be a safe way to back away from a strategic policy based on deterrence and a sure road to peace and stability. This claim, however, ignored the fundamental dilemmas of verifiability. A bilateral nuclear-weapons freeze in the absence of the liberalization reforms Sakharov called for could not be verified any more than could the freeze on chemical and biological weapons. It would have raised the same unanswerable questions concerning what amounts and forms of evidence, and from what sources, would constitute absolute proof of non-compliance in the absence of free and open communications. It would have done nothing but assure that in the U.S., where treaty violations cannot be for long hidden from a free press, the credibility of the nuclear deterrent would have been quickly eroded without any commensurate erosion in the Soviet Union, where testing and production could have continued free of public scrutiny and question, especially if on-site treaty monitoring was not allowed.

The argument has nothing at all to do with the numbers of weapons on either side. In this context questions of numerical parity or superiority, total throw weight or megatonnage, are meaningless and irrelevant. The key point that Freeze activists seemed continually to miss throughout the Cold War, is that if linkage is ignored there could be no assurances on the adequate monitoring or verification of a test ban treaty. There was not a single responsible scientist who could argue at the time that the credibility of a strategic deterrent based on new and modern designs could be maintained in the absence of testing.[4] Furthermore, in the absence of linkage, such efforts as the "No First Use" initiative pushed by various peace groups and the various local and state ballot referenda to establish "Nuclear Free Zones" became meaningless exercises in semantics.

In contrast to the various local initiatives of various peace and Nuclear Freeze groups, the Los Angeles County Board of Supervisors agreed in January 1984 to place before the county's voters on the June primary ballot a referendum on arms control that was based very clearly on linkage. The ballot measure read as follows:

Shall the Los Angeles County Board of Supervisors transmit to the leaders of the United States and the Soviet Union a communication stating that the risk of nuclear war between the United States and the Soviet Union can be reduced if all people have the ability to express their opinions freely and without fear on world issues, including a nation's arms policies; therefore, the people of Los Angeles County urge all nations that signed the Helsinki International Accords on Human Rights to observe the Accords' provisions on freedom of speech, religion, press, assembly, and emigration for all their citizens?[5]

This measure had widespread support from Soviet Jewry activists and other human rights groups, but was generally opposed by the peace activists. Jo Sedeita, Chairwoman of Southern Californians for a Nuclear Weapons Freeze testified against the ballot proposal before the Board of Supervisors, saying that the issues of arms control and human rights should be separated. A member of the Thursday Night anti-nuclear group of Santa Monica publicly expressed similar sentiments, adding that "trying to put pressure on the Soviet Union in this way is counterproductive to increasing our security from nuclear war,"[6] a theme remarkably similar to that expressed in the Radio Moscow broadcast of two months before, and mentioned in Chapter 3. This is not to suggest that both the Soviet Union and U.S. peace groups had the same motivations in opposing linkage. The consequence of their opposition, however, was the same for both: the need to detract attention from Soviet human rights abuses. Thus, Ms Sedeita failed to inform the Board of Supervisors that only a few days before she testified at the Board meeting, Olga Medvedkova, leader of the unofficial Soviet peace movement, was arrested in Moscow along with two Jewish refuseniks. Her "crime" was to remind the Soviet government publicly of former President Brezhnev's proposal on no first use of nuclear weapons. Ms Sedeita also neglected to tell the Supervisors that had Ms Medvedkova been able to speak to them she would have undoubtedly argued in favor of the ballot measure. The Supervisors would have understood that Ms Medvedkova's experiences as a peace activist were much more relevant to the arms control issue than were Ms Sedeita's.

Nevertheless, the U.S. peace movement did make some important contributions to ending the Cold War. Their principal contribution, perhaps, was simply bringing the issue of nuclear arms control and strategic deterrence into the public consciousness. They tapped into a huge reservoir of public concern and fear about the Cold War drifting toward actual nuclear war. Indeed, then-Congressman Albert Gore, on a trip back to his district in Tennessee in 1980, was surprised to hear constituents tell him their view that nuclear war was inevitable. It was reportedly these revelations that launched him on his quest to learn everything he could about arms control and to take a leading role in the Congress in this area. The peace movement played a major role in putting the issue of nuclear arms control on the public agenda.[7]

Additionally, the peace movement helped to change the fundamental thinking about nuclear weapons in the public mind. More and more people throughout the 1980s began to question the morality of the U.S. policy of strategic deterrence as they realized that, as Sakharov had pointed out 20 years before, nuclear weapons were unusable and nuclear war unwinnable. The consequences of a superpower nuclear war would be so devastating to both sides that basing the country's defense on planning for such a war seemed to growing numbers of people illogical as well as immoral. A group of medical doctors, Physicians for Social Responsibility, published several scholarly studies[8] detailing the medical and public health effects of nuclear war. Their approach was not unlike that taken earlier by Sakharov, who came to his activism in nuclear arms control by his concern about the devastating worldwide effects – medical, environmental, social – of a superpower nuclear war. The analyses of Physicians for Social Responsibility seemed to pick up where Sakharov's had left off. Their reasoned pronouncements received wide press coverage in all the Western countries, and they attracted support from large numbers of physical scientists as well as medical scientists and biologists. In some sense, the work of the scientists in studying the consequences of nuclear war, and publicizing their findings, was the spark that initiated the U.S. Nuclear Freeze movement.

The scientists' calling public attention to the terrible consequences of nuclear war also captured the attention of those in government, career civil servants as well as political appointees and elected public officials. Most people in government, like most Americans, were neither on the left nor the right in the arms control debates. They wanted neither a complete abandonment of nuclear weapons, as did the left, nor a complete rejection of arms control, as did the right. Those who thought about the issue at all mainly wanted the Cold War conflict to be managed in such a way as to prevent it from erupting into direct military conflict between the U.S. and the U.S.S.R. This meant using arms control as a means to limit and contain the Soviet expansion of Communism and its military threat. In other words, arms control was seen as a tool of U.S. foreign policy. The central doctrine of U.S. foreign policy throughout nearly the entire duration of the Cold War was the doctrine of "containment." As first promulgated by George Kennan and a few other young State Department employees during the administration of President Harry Truman in the early days of the Cold War, containment meant that the U.S. would oppose, by force, if necessary, any expansion of Communist rule outside the borders of the U.S.S.R.[9] If this meant, as it often did, forging alliances with unsavory characters in third-world countries in order to defeat Communist insurgencies in those countries, then so be it. In this view, there was nothing more important to the security interests of the U.S. than stopping and containing the worldwide spread of Communism. It was the doctrine of containment, of course, that was used to justify U.S. military involvement in Vietnam, in Central America, in Angola, and even the U.S.-backed overthrow and assassination of the elected President of Chile in 1973. As Cold War historian John Lewis Gaddis has pointed out, though, there would have been no U.S. policy of containment had there not been something that truly needed containing.[10]

The right, of course, had a problem with the doctrine of containment because it was, in their view, not ambitious enough. The containment advocates in government wanted only to stop the spread of Communism, not eliminate it entirely. Thus, they did not seem to the right to be displaying sufficient faith either in the ability of the U.S. to defeat Communism or in the Russian people to free themselves from the tyranny under which they lived. The containment practitioners, on the other hand, thought that they were the practical ones, advancing a foreign policy goal for the U.S. that was not only achievable but that was more likely to avoid superpower war for the long term.

The left also had a problem with the doctrine of containment.

To the left, the U.S. government's willingness to rely on nuclear deterrence as long as it seemed to advance the cause of containment of Communism was morally unacceptable.

There was one area, however, in which the containment advocates were in full agreement with both the right and the left: they also opposed linkage of arms control with human rights and democratization. Their opposition to linkage was for reasons quite similar to those of the left: for them, arms control, in so far as it advanced the foreign policy goals of the U.S. – containment of Communism – was of such paramount importance that no other considerations should be allowed to get in the way. Kennan himself on many occasions expressed his clear opposition to allowing U.S. foreign policy to be guided by human rights concerns. Henry Kissinger, who was Assistant to the President for National Security Affairs (1969 - 1975) and U.S. Secretary of State (1973 - 1975) under Presidents Richard Nixon and Gerald Ford, expressed the same sentiments[11].

Kennan and Kissinger were not the only anti-linkage people who exercised significant influence on U.S. foreign policy during the course of the Cold War. The doctrine of containment became more or less an article of faith in policy circles, both Democratic and Republican. Opposition to linkage was strongly embedded in this doctrine. Hence, Jackson-Vanik aside, the U.S. generally did not follow a consistent policy of linkage. Only the scientists argued for it. And it was finally in the Soviet Union, not in the U.S., that the scientists' argument was heard and acted upon.

1. "Soviet Jews and Nuclear Disarmament," by Joseph Ribakoff, *Israel Today*, December 13, 1983.
2. *Physics Today*, Vol. 33/4, p. 81, April 1980.
3. The Helsinki Commission reports can be found at www.csce.gov.
4. "Debate on a Comprehensive Nuclear Weapons Test Ban," by Hugh E. DeWitt and Robert B. Barker, *Physics Today*, Vol.

36/8, August 1983.
5. *Los Angeles Times,* Wednesday, January 4, 1984.
6. *Santa Monica Evening Outlook,* Wednesday, January 4, 1984.
7. "How We Ended the Cold War", by John Tirman, *The Nation,* p. 13, November 1, 1999.
8. The reports of the Physicians for Social Responsibility can be found at www.psr.org.
9. See, for example, "Reflections: Breaking the Spell," by George Kennan, *The New Yorker*, October 3, 1983.
10. *We Now Know: Rethinking Cold War History* by John Lewis Gaddis, Oxford University Press, Oxford, U.K., 1997.
11. In "Challenges to the West in the 1980s," a speech Kissinger gave to the 5th CSIS (Center for Strategic and International Studies) Quadrangular Conference, Georgetown University, Washington, D.C. on September 20, 1982, and reprinted in his book *Observations: Selected Speeches and Essays, 1982 - 1984* (Little Brown and Co., Boston, 1985), he says "I suspect that if the various arms control proposals are analyzed in detail, we would find that they are much too driven by the need to deal with immediate pressure groups (author's note: read 'human rights groups') and much too little geared to the security situation we foresee in the middle eighties." Kissinger also argued that the failures of the containment policy stemmed from the failures of Western leaders to recognize that the West was militarily strongest vis-a-vis the Soviet Union at the very beginning of the Cold War (the end of World War II), when the U.S. had a nuclear weapons monopoly and the Soviet Union was in economic ruin; the U.S. should have then, according to Kissinger, recognized the need to conduct negotiations and diplomacy from its position of strength. Thus, Kissinger was not opposed to a policy of containment, but he saw the failure of the U.S. to link containment to a larger foreign policy strategy and diplomacy as a mistake. He also saw as

a mistake the linkage of arms control, in so far as it served the objectives of containment, with human rights concerns. Kissinger thus had the same position on linkage as did the left, but, of course, for very different reasons.

Chapter 7

The Scientists' Boycott

Benjamin Franklin's discovery of positive and negative electric charges as the carriers of electricity, as mentioned in Chapter 2, was a profound discovery, for it opened up a line of worldwide scientific investigation that ultimately led to technological innovation becoming the driver of the world's economies and transforming the industrial revolution. Experiments and mathematical studies conducted in the nineteenth century by André Marie Ampère in France, by Karl Friedrich Gauss in Germany, by Michael Faraday in England, and by Joseph Henry in the U.S. established the basic laws of electricity and magnetism. Ampère showed that an electric current flowing in a coiled wire acts just like a magnet, establishing that magnetism arises from the motion of electric charges. He also showed that electric currents flowing in parallel wires exert a force on each other, establishing the existence of a magnetic force that complements the electric force. Gauss developed the law of conservation of charge, and helped establish the concept of an electric "field" as the "medium" that transmits the electric force.

This "field" idea was an important advance on the concept of "action at a distance" that Newton used to explain the behavior of the gravitational force. Nineteenth-century scientists began to reject the Newtonian notion of "action at a distance" in favor of the idea that a force requires some "medium," a "field," to transmit its effects. Faraday and Henry, in a number of brilliant experiments performed independently on opposite sides of the Atlantic Ocean, further refined and clarified the behavior of electric and magnetic fields. They established the basic laws of electric circuits, discovering electrical resistance and magnetic induction in the process.

Scientists' work on understanding the nature of electricity and magnetism had enormous practical benefits. Not the least of these was the invention of the electric motor and its inverse, the electric generator. The electric motor, of course, ultimately replaced the steam engine as the main driver of the industrial revolution. Industrialization enabled rapid economic growth and expansion, creation of new jobs, and a general rise in the level of people's prosperity and well-being. To be sure, many challenges were presented to democratic government in making this economic prosperity available to the entire population. Innovation could only flourish in a democratic culture, a lesson that seems to have to be relearned by every successive generation. Nonetheless, it is true that the average person in democratic countries today lives better – as measured by life span, general health, quality of housing, living standards, and many other measures – than most aristocratic and royal families of pre-Renaissance Europe. All this is owing in large measure to science and technology. The electric motor was only the beginning of a vast transformation in the way people live and the way they understand the world in which they live.

James Clark Maxwell, an English physicist, built upon the work of Ampère, Gauss, Faraday, Henry, and others. He derived the basic mathematical laws of electromagnetism. He showed that electric and magnetic fields can be combined mathematically to create an electromagnetic wave motion, and that the electromagnetic wave, whatever its wavelength, moves at the speed of light. Thus, Maxwell predicted the existence of both radio waves (electromagnetic waves of very long wavelength) and x-rays (electromagnetic waves of very short wavelength) before either was even discovered. He also showed that ordinary light is itself an electromagnetic wave. Maxwell's work led directly to the invention of radio, and became the foundation for all of modern electronics.

Maxwell, however, like all nineteenth-century scientists,

believed in the existence of an invisible "ether" that pervaded all space. The ether was assumed to be the medium in which electromagnetic waves propagate. It was thought that all waves had to have some substance in which to move, much like acoustic, or sound, waves need air or some other material in which to propagate. As we all know, if there is no air (or some other substance) there is no sound. (Rocket motor sounds cannot, of course, be heard in the vacuum of outer space, notwithstanding how they are presented in such science-fiction movies as *Star Wars*.) The American physicist Albert Michaelson thought of a way to detect the existence of the ether. He figured that a light beam traveling along the surface of the earth in the same direction that the earth travels through the background ether would appear to be going faster with respect to the background ether than a light beam traveling in the opposite direction, against the ether. This conclusion follows directly from the basic laws of motion discovered by Galileo. Imagine, for example, a person throwing a ball on a train as observed from inside and outside the train. To the observer sitting in the train, and moving along at the speed of the train, it would look like a ball tossed in the direction of the train's motion would be going the same speed as one tossed in the opposite direction. An observer on the ground, however, watching the train and the ball thrower go by, would observe the ball tossed in the forward direction moving faster than the one tossed backwards.

Michaelson invented a "wave interference" device, an "interferometer," to be able to measure these very small differences in light speed in the frame of reference of the stationary ether. To his astonishment, and to the astonishment of all the world's scientists at the time, Michaelson discovered that the speed of light is constant regardless of its direction of propagation. Not only is there no ether, but electromagnetic waves do not seem to obey the elemental laws of motion that Galileo had discovered and formulated a couple of centuries before! This astonishing discovery opened the way for Albert Einstein's theories of relativity. The Theory of Special Relativity begins from the postulate that the speed of light is a constant in all frames of reference moving at different velocities with respect to one another. Thus, it does not matter in what frame of reference the measurement of light speed is made; one will always obtain the same result. This means that the speed of light is an upper limit to how fast anything can move. Einstein's work thus changed all our basic concepts about space, time, motion, and their relationships to one another. Human understanding of the cosmos has been altered by these concepts of relativity theory in a more profound and fundamental way than the changes in thinking brought about by Galileo's original discoveries with his telescope.

The efforts to understand electromagnetic waves led to many more astonishing discoveries that have transformed human society. One of the early mysteries about electromagnetic waves concerned their interaction with ordinary matter. European scientists had discovered that heated material emits a "spectrum" of electromagnetic radiation, i.e., waves of different wavelengths, and that this thermal spectrum is characterized only by the material's temperature. All attempts failed, however, to derive this thermal spectrum mathematically from the known laws of the thermal motion of the atoms and molecules of the material that is emitting the waves. Finally, the German physicist Max Planck found that he could derive the measured thermal spectrum if he postulated that the energies of the vibrating atoms were constrained to have only discrete values, only multiples of a very small "quantum" of energy. Planck's quantum hypothesis was initially only that – a hypothesis. It soon took on the characteristics of a fact of nature as it began to explain a lot more mysteries concerning the interactions of electromagnetic radiation with atoms.

By assuming that the electromagnetic field itself is quantized, rather than the vibration energies of the radiating atoms, Einstein was able to explain the so-called "photoelectric effect," in which shining light on a material can cause a current to flow in the material, but only if the wavelength of the light is sufficiently short. For this work, Einstein won the Nobel Prize in Physics (not for relativity theory, as is commonly believed, even though the paper on the Theory of Special Relativity appeared in the very same issue of the German physics journal as his paper on the photoelectric effect).

Planck's quantum postulate was adapted to solve the biggest puzzle of all in early twentieth-century science: why atoms are stable. The problem was this: if electrons are spinning around the nucleus of atoms, why do they not radiate away all their energy, since we already know that moving electric charges produce electromagnetic radiation? The answer to this puzzle emerged from the realization that fundamental sub-atomic particles like electrons really behave as waves, with their wavelength dependent on their velocity. This hypothesis was put forward by a young French scientist, Louis de Broglie, in his Ph.D. thesis, for which he won the Nobel Prize. Then, the Danish physicist Neils Bohr showed that these electron "matter waves" could only exist in certain "quantized" orbits, namely those orbits in which exactly a whole number of wavelengths can fit around the circumference of the orbit. Only in these orbits is the electron stable, much like stable standing waves on a guitar string. Bohr's discovery was the key to explaining the atomic structure of all the elements and the characteristic spectra of electromagnetic radiation that each one emits. The development of "wave mechanics," or quantum theory, following Bohr's discovery, became the foundation of all modern technological development, from lasers to computer chips.

Nineteenth-century and early twentieth-century chemists also made great advances in understanding that materially changed human life. The new understandings of atomic structure and quantum theory led to understandings of the basics of chemical bonding, how atoms of different elements "share" electrons to bind together into molecules and compounds. Researchers also figured out the basic mechanisms and rates of chemical reactions. Out of this research emerged the technology for creating new materials, from plastics to petroleum fuels to synthetic fibers.

Scientific advances in biology and medicine were no less impressive. Perhaps the most profound and significant scientific discovery of the nineteenth century was the discovery that infectious disease is caused by living organisms. The "germ theory" of disease was highly controversial in the nineteenth century, but its confirmation by long years of painstaking scientific research directly affected the life of every human being as it led to the development of antibiotic medicines.

It is easy to understand how scientists, from their study of the historical developments in science, and the impact of these developments on human society, are, in general, believers in progress. Although few scientists embark on their careers specifically to work for the betterment of humanity, many of them accept the notion that increasing human understanding of the natural world is a net good, and that society will benefit in the long term from their work. They also clearly understand that science itself progresses only through free and open international cooperation and communication with other scientists. Maxwell's development of electromagnetic theory, for example, would have been impossible without his having ready access to the results of Ampère in France, Gauss in Germany, Henry in the U.S., as well as Faraday in his native England. Open communication of research results is such a fundamental part of the scientific enterprise, and the scientific enterprise such a fundamental contributor to the betterment of human society in the thinking of scientists, that any political interference with such communication is automatically opposed. Scientists, in fact, take it as a direct threat to the fundamental integrity of science when there is any political interference in the free and open communication on which the enterprise depends.

Thus, one of the biggest mistakes the Soviet government made during the Cold War was underestimating the reaction of the worldwide scientific community to their clumsy interference in the scientific enterprise. This interference took several forms.

The first and perhaps most direct form of Soviet interference in science was the Soviet government's refusal to issue entry visas to particular foreign scientists to attend scientific meetings and conferences in the U.S.S.R. Israeli scientists, in particular, were routinely denied visas to enter the U.S.S.R. Some American scientists, particularly those who were known to Soviet authorities as having engaged in proscribed activities on previous visits – like visiting refusenik scientists without permission – were also denied visas. Others had their visas revoked and were forcibly expelled from the country after visiting refusenik scientists while in the country to attend an "official" conference.[1]

It was also a common practice for Soviet authorities to place political restrictions on which Soviet scientists would be allowed to attend scientific conferences abroad. All foreign travel was under the direct control of the KGB, the Soviet secret police. Refusenik scientists, of course, were barred from leaving the country for scientific meetings. Many well-known Soviet scientists who were not refuseniks were also not allowed for political reasons to travel abroad. Those who were allowed out invariably came with a KGB escort. Another insidious and widespread practice was for the KGB to make a last-minute substitution, sending in place of a recognized scientific expert some lesser-known individual who had been properly vetted by the KGB.[2]

These were not occasional isolated incidents, but routine happenings, so much so that it became nearly impossible for

American or West European conference organizers to count on Soviet participation in any scientific conference. The organization of a scientific conference, particularly an international one, is a major undertaking. The enormous effort involved in organizing such a conference was routinely sabotaged by what became a common Soviet practice of submitting a large number of abstracts for intended conference speakers, only to have less than half of the scheduled speakers actually show up. Worse, often Communist party hacks were sent as stand-ins for scheduled speakers, or would show up unannounced and demand time on the conference schedule.

Of course, the most severe interference in scientific communication came with the U.S.S.R.'s scientific isolation of refusenik and dissident scientists like Andrei Sakharov. As already discussed in Chapter 3, refusenik and dissident scientists typically lost their jobs. With the lost job they also lost all access to research facilities, laboratories, even libraries. Some refusenik scientists even had their university degrees revoked,[3] in a clear attempt not only to isolate them but to delegitimize and humiliate them as well. As if this action was not enough, the government even went so far as to erase the person's very existence by removing all references to that person's work from published scientific papers and citation indexes.[4]

Not only were there continual efforts to inhibit the free flow of scientists to and from scientific conferences and laboratories, there was strict control of information flow as well. Scientific journals sent to the U.S.S.R. were censored. Articles, news reports, and letters published in major scientific journals like *Science* and *Nature* that touched upon human rights, scientific freedom issues, foreign policy, arms control technology, and a large array of even technical issues were completely excised from the journals before they were sent on to the intended recipients in the U.S.S.R. Often, the journals did not reach their intended recipients.

The initial response of American scientists to these attempts to isolate their Soviet scientific colleagues was to take individual actions to alleviate their colleagues' isolation from scientific communication. More letters were sent describing scientific research findings, more phone calls were made. Soviet authorities, however, stepped up their own attempts at the interception of mail and telephone calls. Letters meant for particular recipients were routinely never delivered. Telephone calls placed to particular individuals often failed to go through. Nonetheless, the volume of mailings and calls did not decrease, because despite the reach of the Soviet bureaucracy it was, if anything, very inefficient, so not all mail and not all calls could be intercepted. American scientists also took to regular mailings of article preprints and personal copies of scientific journals.

These individual efforts had limited effectiveness because, of course, only a fraction of the communications that were sent were actually received. More organized efforts evolved as the professional societies became involved. Many of the large professional societies of scientists formed human rights committees as "permanent" standing committees of the society, chartered to work on behalf of scientific freedom issues worldwide. This typically meant recommending and carrying through strategies to alleviate the plight of Soviet scientists who worked, or had worked, in the same scientific discipline.

One of the first such organized efforts was undertaken by the Association of Computing Machinery (ACM), a professional organization of computer scientists. The ACM was compelled to action particularly by the case of fellow computer scientist Anatoly Shcharansky, who, with physicists Yuri Orlov and Andrei Sakharov, had founded the Moscow Helsinki Watch Group to monitor and report on the U.S.S.R.'s compliance with the human rights provisions of the Helsinki Accords (see Chapter 3). Shcharansky had been arrested, tried, and sentenced originally to three years in prison and 10 years in a Siberian labor camp. During the initial part of his sentence at the Perm Labor Camp, he was repeatedly accused of trivial violations of camp rules and put in an isolation cell for long periods of time and reportedly denied adequate food and medical care. His original sentence was revised to six years in prison and seven in a labor camp, and in 1982 he was transferred to the KGB's Chistopol prison in Moscow.

Partly in response to Shcharansky's plight, the ACM formed a Committee on Scientific Freedom and Human Rights. The Committee publicized particular cases of government interference with scientific freedom, and organized letter-writing campaigns to the affected scientists, as well as letters of protest to Soviet government authorities. Like all committees, they also issued regular reports on their activities. The ACM's Committee on Scientific Freedom and Human Rights issued yearly reports, which were published in their professional journal, *Communications of the ACM*, starting in 1981. These reports contained details on all the particular cases on which the Committee had worked that year. The report for 1982,[5] for example, has a paragraph on each of 53 Soviet computer scientists, in addition to a paragraph on each of 15 other computer scientists in six other countries. The 1982 report also lists 27 additional Soviet computer scientists whose status had not changed since the previous year's report. By the time of the 1984 report, the list of computer scientists denied their scientific freedom had grown to 45 in Poland and 135 in the Soviet Union.[6] Shcharansky was still on the list and still in Chistopol Prison. He had been placed that year under the prison's "strict regime", which means his food rations and his exercise time outside his cell were both reduced.

It was not only the computer scientists who were active on behalf of their Soviet (and other) colleagues. The American Physical Society, the largest professional organization of physicists, formed the Committee on the International Freedom of Scientists (CIFS). This committee, like its ACM counterpart, also issued reports, took up the cases of beleaguered colleagues abroad, and organized letter-writing campaigns and other member activities. CIFS also recommended to the Society individual refusenik or dissident scientists who should receive complimentary subscriptions to Society-published scientific journals.[7]

The American Chemical Society, the largest single-discipline scientific organization in the world – it had over 100,000 members during this period of time – organized a subcommittee of the Committee on International Activities in 1978 to deal with scientific freedom and human rights issues. The Society also took action on behalf of particular cases that were brought to the attention of the Board of Directors by individual members of the Society. The American Statistical Association (ASA), a professional organization of mathematicians, also had a Committee on Scientific Freedom and Human Rights. In fact, the ASA received a two-year Ford Foundation grant in 1985 to study ways in which statistics could be used to evaluate and improve the measurements of countries' human rights performance.[8]

The world's largest scientific organization, the American Association for the Advancement of Science, chartered its Committee on Scientific Freedom and Responsibility (CSFR) in April 1976. The CSFR was a large organization in itself, consisting of five subcommittees covering a broad range of issues, from "Scientific Freedom in Foreign Countries" to "Social Responsibilities of the Scientist." Like their counterpart committees of other professional associations, the CSFR collected information on individual case histories and recommended collective action to the membership and leadership of the association. They also published annual reports summarizing their activities. The CSFR established the Clearinghouse on Science and Human Rights to serve as a central collection point for information on scientists in foreign countries whose human rights and/or scientific freedoms had been violated. The Clearinghouse defined these rights to include freedom of expression and publication; free access to education and employment; freedom of assembly and association; and freedom of movement and residence, including the freedom to attend international scientific meetings.[9] The Clearinghouse established formal contacts with 33 other professional societies and associations, including the American Physical Society, the American Chemical Society, the American Statistical Association, and the Association for Computing Machinery. Life scientists were represented as well, through such organizations as the American Society of Zoologists and the American Medical Association.

Another activity that was pioneered by the AAAS was the organization of human rights sessions at their scientific Annual Meetings. These meetings provided an opportunity for more in-depth discussions among scientists of the particular issues raised by human rights and scientific freedom concerns. The idea of special sessions and symposia at scientific meetings caught on; by the mid-1980s many other scientific societies like the American Physical Society were holding such sessions at their Annual Meetings.

In addition to the already-existing scientific societies, new organizations of scientists sprang up that were devoted entirely to the struggle for scientific freedom of oppressed scientists. The most prominent of these organizations was the Committee of Concerned Scientists, headquartered in New York. Individual scientists became members of CCS by payment of a yearly membership contribution, and many scientists from a wide range of disciplines – including me – participated in CCS activities. Much like the AAAS Clearinghouse, CCS served as a central collection point for reliable information on particular cases in the Soviet Union and elsewhere. The Committee established a large network of reliable contacts in the refusenik and dissident communities in the U.S.S.R., and thus had access to much first-hand information on developments there in particular cases.

Certainly, a key role played by the professional societies and the CCS was obtaining publicity and press attention for particular cases, as well as for the overall issue of Soviet suppression of scientific freedom. Thus, not only were human rights symposia highlighted in press releases about scientific meetings, but scientific meetings became the venue for more direct action taken on behalf of particular oppressed scientists. Such action often attracted press attention. Collection of signatures on petitions became a particular focal point of activity. Petitions were usually addressed to Soviet authorities on behalf of one or more individuals whose plight was being addressed by the petition. Petition drives at American meetings usually went smoothly. This was not always the case at international meetings, particularly meetings at which there was an official Soviet presence. At one such meeting, for example, the author was threatened with arrest and expulsion from the meeting place because of refusing to stop petition activities after the Soviet government lodged a formal complaint with the host government.[10]

Here is what happened in this particular incident. In early 1979 I had received an invitation to present my research at a meeting of the International Astronomical Union (IAU) in Montreal in August. I contacted CCS before going to Montreal, and agreed to work cooperatively with two Canadian colleagues and one American colleague who were also going to be attending the conference in Montreal to collect petition signatures from other conference participants in support of two Soviet colleagues. Both Soviet scientists had been fired from their positions after they had applied to emigrate, and were then in the untenable position of being allowed neither to emigrate nor to continue their scientific work. One of these refusenik scientists, Vladimir Dashevsky, was expected to be facing possible charges of "parasitism" and its attendant prison sentence, since it was then against Soviet law for a Soviet citizen to be unemployed.

After my arrival in Montreal, and after I made contact with the three colleagues with whom I was to work on the petition signature collection effort, I then arranged with a local friend to obtain a small folding table that I could set up near the entrance to the cafeteria, which was located near the meeting rooms in the building where the conference was being held. Copies of the petition were placed on the table so that conference participants could see it and sign it if they wished to do so. My two Canadian colleagues told me that they had been assured by the conference organizers that, although they could not officially endorse our efforts, they also would not interfere. The four of us agreed on a schedule of taking turns at the table during breaks in the conference schedule, or in between sessions, or during the lunch break, all times when more people would be passing by the table.

Everything went fine for the first couple of days, and we four collected lots of petition signatures. Then on the second or third day, when I was at the table by myself, I was approached by two people, one of whom identified himself as representing the Canadian National Organizing Committee. He informed me that an official protest of our petition effort had been lodged by the Soviet government through the Soviet Consul General in Montreal, and that if I did not immediately cease my activities inside the building that the Soviet conference participants would walk out, and that the police would be called to eject me.

Upon hearing that Soviet conference participants threatened to walk out, I remember responding with a humorous remark about this being a good thing because then the lines would be shorter in the cafeteria. The Canadian officials, however, were not amused.

Psychiatrists and psychologists tell us that a situation of imminent danger triggers a "fight or flight" response in the person facing the danger or threat of danger. For me, however, I have often used humor as a coping mechanism to control the "fight or flight" response, because I believe, perhaps wrongly, that it can buy me more time to consider a more rational response to the threat, especially if the danger is not imminent, like in this particular encounter. I was not meaning to be flippant or defiant, but I can understand now, in retrospect, how the two Canadians could have interpreted my remark as flippancy or defiance. In any case, I was so unnerved by this encounter – as well as enraged that Canadian officials had allowed themselves to be threatened and intimidated by the Soviet government – that I decided to write a letter about this incident to all the members of the IAU Executive Committee. I wrote the letter the next day. Then, the following day (that is, two days after the threat to have the police throw me out), before I sent the letter and while I was at the table again, this time with one of my Canadian colleagues, I was approached by another member of the Organizing Committee. He offered me an apology, telling me that the person who threatened me two days before did not speak with full authority and had been admonished. I accepted the apology and agreed not to proceed with the letter.

Reflecting back on this incident after I had returned to my lab in the U. S., I decided that there were important issues raised by this incident at the IAU meeting that I sent a revised version of the letter I had written to the IAU Executive Committee for publication in the main trade magazine of physics. It was published in February 1980, along with a response letter from the Canadian official who originally threatened to have me ejected from the meeting.[10]

Even as early as the mid to late 1970s, many scientists felt that all these actions – letters, petitions, personal appeals, articles in professional journals and the press, visits to refusenik and dissident scientists, even collective actions by the professional societies – were simply not enough, especially since little seemed to change in the plight of the affected Soviet scientists as a result of these actions. Many scientists argued for more forceful action. The idea of a boycott arose in many places seemingly simultaneously. After the imprisoning of Shcharansky, many scientists had made individual decisions to stop all cooperation with Soviet scientists to protest Shcharansky's treatment. This was an enormous step for scientists to take, because we have already seen how scientists hold fast to the idea that free and open scientific communication and cooperation is fundamental to the scientific enterprise. Under normal circumstances, a scientific boycott would be a contradiction for scientists, and would not generally be supported. These, however, were anything but normal circumstances. Just as human rights activists argued for support of the Jackson-Vanik Amendment to the Trade Act of 1974 (see Chapter 5), many scientists now argued that "business as usual" was no longer possible.

The idea of an organized boycott started with groups of American scientists pledging themselves to boycott a specific upcoming international meeting to be held in the Soviet Union. One of the first such actions was an American boycott of the International Congress of Genetics in Moscow in August 1978. A group of about a dozen American geneticists, including two Nobel laureates, called for the boycott. The boycott organizers acknowledged that some scientists would choose to go to the meeting anyway; those who chose to go were urged to express their concerns about their imprisoned colleagues directly to Soviet conference participants, and even to visit refusenik scientists who found themselves scientifically isolated.

The idea of a more general, massive, worldwide scientific boycott of the Soviet Union originated with a group of scientists at the University of California Berkeley and the Lawrence Berkeley Laboratory (now called the Lawrence Berkeley National Laboratory). The group, originally calling themselves Scientists for Orlov and Shcharansky (SOS) was created during the period of the trials of Yuri Orlov and Anatoly Shcharansky in the summer of 1978. Berkeley Lab physicist Morris Pripstein was the first leader and organizer of the SOS group. The SOS objective was for individual scientists, acting independently, to pledge themselves to restrict personal scientific cooperation with the Soviet Union, and thereby deprive the U.S.S.R. of some of the benefits of American science and technology. Such restrictions were meant to include refusal to attend scientific meetings in the Soviet Union, even if invited, refusal to invite Soviet scientists to visit their own laboratories, refusal to respond to reprint and preprint requests, withholding of any scientific communications with Soviet scientists, and lobbying against any expansion of official scientific exchange programs. Within a few months, SOS had more than 2,400 American scientists who pledged their support. This number included 13 Nobel laureates, 113 members of the U.S. National Academy of Sciences, and presidents (past and then-current) of 20 scientific societies. These scientists signed a pledge "to withhold all personal cooperation with the Soviet Union until Orlov and Shcharansky are released." Companion efforts in France and Australia netted about an additional 1000 pledges of support for the boycott.

This may seem today like not a lot of signatories to a boycott pledge of this scope, but keep in mind that this was done in the days before the internet, before email, before cell phones, before social media. Communication between scientists was done entirely by regular mail and dial telephones. Getting a scientific article published in a refereed science journal was a many-months-long process – sometimes a year or more – as drafts went back and forth between the author, the journal editor, and the referees by regular mail. To speed up the process, the original organizers of the boycott reached out to scientists at other institutes, laboratories, and universities to be recruiters of signature gatherers in their own organizations. In this way, many of the scientists who were asked to sign, or who asked to sign, also were asked to approach others to sign. For example, I obtained copies of the pledge directly from Morris Pripstein, the original boycott organizer at Lawrence Berkeley Lab, and solicited signatories at my lab.

After Andrei Sakharov was arrested and then sent into internal exile to the closed city of Gorky in January 1980 (see Chapter 4), SOS was renamed Scientists for Sakharov, Orlov and Shcharansky, retaining the same initials. Then, SOS put out a call for a complete moratorium on all scientific cooperation with the Soviet Union, and sought a much larger number of signatures. This was the moratorium pledge scientists were asked to sign:

> *To protest the human rights violations by the Soviet Union in the cases of Sakharov, Orlov, and Shcharansky, we, the undersigned scientists and engineers, pledge a moratorium on professional cooperation with the Soviet scientific community for a period beginning 12 May 1980, the anniversary of the founding of the Moscow Helsinki Watch Group, and ending at the completion of the November 1980 Madrid Conference to monitor the Helsinki Accords. During this period, we will not visit the Soviet Union or welcome Soviet scientists and engineers to our laboratories.*

At simultaneous press conferences in Washington, Paris, London, and Geneva on October 16, 1980 – just as delegates to the Conference on Security and Cooperation in Europe were arriving in Madrid for the meeting to review compliance with the 1975 Helsinki Accords – SOS announced that 7900 scientists and engineers from 44 nations had signed the moratorium pledge. This number included 32 Nobel laureates, 187 members of the U.S. National Academy of Sciences, 82 Fellows of the Royal Society of the U.K., and a number of members of the French and Italian Academies. Many of these scientists had previously been active in promoting scientific exchanges with the Soviet Union.

Even though the original moratorium pledge was in effect for a limited period of time, it was clear that suspension of the moratorium was dependent on the outcome of the Madrid meeting. If the Soviet Union could not be compelled to comply with Basket III – the human rights provisions – of the Helsinki Accords, then the boycott would continue. One very important aspect of this boycott was that it was, unlike the trade moratorium of the Jackson-Vanik Amendment, entirely a grass-roots effort. The scientists' boycott was conceived, organized, and carried out entirely by individual scientists acting collectively and independently of their governments. This is what made it so powerful, and an action that the Soviet Union could not ignore.

Of course, nothing much changed after the Madrid meeting. Sakharov remained in exile. Orlov and Shcharansky remained in labor camps in Siberia. If anything, their situations got worse. There were more arrests, a growing number of refuseniks, more crackdowns on dissent, a general deterioration of East-West relations in the wake of the Soviet invasion of Afghanistan, President Reagan's belligerent rhetoric, and propaganda wars over the American Strategic Defense Initiative and arms control issues (see Chapters 5 and 6). The boycott, of course, continued, and even expanded in scope and support. It is important to emphasize once again that agreeing to a boycott was a difficult step for scientists to take, because it went counter to scientists' natural inclination to promote free and open scientific communication. The argument that boycott organizers used, however, was that the boycott itself would be both appropriate and effective in bringing about freedom of scientific communication.

The SOS press conferences on October 16, 1980, received extensive worldwide press coverage. Stories about the boycott were carried by the three major commercial news networks in the U.S. – ABC, CBS, and NBC – in both TV and radio broadcasts. The major newspapers, like the *New York Times* and the *Washington Post,* carried the news, too. The *Associated Press* and *United Press International* wire services carried stories about the boycott, which were reprinted in numerous newspapers all over the world. The news also prompted newspaper editorials. The *Washington Post* editorial of October 19, 1980, was typical. The *Post* proclaimed that the boycott protests "have a special force precisely because they are not linked to any government's policy, but represent the acts of individuals speaking for themselves. The scientists' protests are the authentic voice of conscience – it is good that Soviet authorities are being compelled to listen to it."

In the immediate aftermath of the SOS boycott, a number of scientific meetings that had already been organized were canceled. An accelerator conference in Dubna, Russia was canceled at one week's notice. The French organizers of a physics conference held in 1980 issued "anti-invitations" to Soviet participants. These were letters stating that while colleagues like Orlov remained in a prison camp it was unfortunately impossible to invite them to the conference.

Not all scientists supported the boycott. There were the inevitable debates over the effectiveness of the scientists' boycott, just as there had been some five years before over the Jackson-Vanik Amendment. The principal arguments against the boycott were that it was ineffective and inappropriate. William Carey, then Publisher of the journal *Science,* published an editorial in the journal on October 24, 1980, in which he expressed alarm over the latest developments in East-West relations. "The position is that we are very nearly out of safety valves as the nuclear superpowers drift toward impasse," wrote Carey. "If scientific responsibility is more than an idle phrase, it requires participation in the pursuit of peace and conflict resolution. The quarantining of Soviet science, however principled, defeats the chances for engaging a concerned and far from impotent cohort of opinion and influence in a dialogue of reason." In other words, Carey was expressing the same concerns expressed by the American left during the debate over Jackson-Vanik (Chapter 6), namely, that provoking the Soviets over human rights issues, no matter how principled, risked exacerbating the tensions that could lead to nuclear war. Thus, the boycott was deemed to be inappropriate because it was directed at the wrong people, i.e., at the "progressive" Soviet scientists who are in a position to work with us to prevent nuclear war.

This criticism was, however, uninformed. Since the Soviet science establishment was designed to be an instrument of state policy it was subject to a bewildering array of authoritarian control. Intellectual freedom was all but nonexistent.

It was the dissident and refusenik scientists who were the real "progressives," and they were all thoroughly isolated from the world of science by their much less progressive Soviet colleagues, and in many cases (like Orlov and Shcharansky) were in prison camps. The boycott was directed at Soviet scientists who either participated actively in the oppression of their colleagues or condoned it by their silence, and this, unfortunately, was the majority of them. As it was put by five American Nobel laureates who supported the boycott, the critics "have misconstrued the basis for our actions and have gravely underestimated the depth and extent of the disaffection of American scientists engendered by the oppressive actions of the Soviet authorities."[11] They argued that the boycott was not only appropriate, it was the necessary response to Soviet actions.

As for the effectiveness of the boycott, different facts were elicited to support different sides of the argument. It is true that neither Sakharov, Orlov, or Shcharansky were released as an immediate and direct consequence of the boycott. It is also true that in many instances the effect of the boycott seemed to be merely a delay of a few months in the receipt of American science journals in Soviet laboratories. On the other hand, it is also true that authoritarian state control of the science enterprise in the Soviet Union meant that, despite the large Soviet money and manpower investment in science, it was a largely unproductive enterprise, dependent in a way many Americans failed to appreciate on Western creativity and initiative. Thus, the boycott represented a much more severe blow to the Soviet Union than a few months' delay in receiving scientific articles. As evidence of this assertion, one need only point to the sharp reaction of the Soviets themselves to the boycott, which included two English-language broadcasts beamed to North America from Moscow, in which Communist Party commentator Valentin Zorin denounced the original boycott (17 March and 19 May 1979), and a long editorial in *Pravda*, the Communist Party newspaper (23 April 1979), signed by five senior members of the Soviet Academy of Sciences.

Surely these officials of the Soviet science establishment were not treating the moratorium as ineffective. Of equal importance, the Soviets had an almost obsessive need to maintain an image of international respectability in science. The psychological damage they risked suffering by the boycott was not to be underestimated. The Soviet fear of such damage turned out to be a source of considerable leverage in effecting fundamental change, as we will explore in the next chapter.

Proponents of the boycott argued that to be truly effective the boycott had to be sustained for the longer term. They recognized that Soviet authorities regarded the boycott as simply a passing event. The Soviets wanted and needed the scientific contacts, but were not willing to alter their behavior to get them as long as they thought that they could simply outlast an initial period of protest and harsh criticisms, after which the boycott would vanish into the background in the greater interests of reducing international tensions and preserving the peace. To boycott proponents this Soviet attitude called for stronger resolve in sustaining it. "Clearly, we must stand firm," said chemist Paul Flory, one of the SOS leaders. "To do otherwise would jeopardize many courageous people, deprive them of hope, and discourage others from trying. Moreover, it would strengthen the hands of the hard-liners, and this would worsen prospects for a resolution of international tensions."

Flory expressed similar views in Congressional testimony, although he did not advocate a complete cessation of formal government-to-government scientific exchange agreements. He did, however, recommend "minimum conditions" to which the Soviets should adhere in such agreements. At a joint hearing of the Commission on Security and Cooperation in Europe, the Subcommittee on International Security and Scientific Affairs, and the Subcommittee on Science, Research and Technology on January 31, 1980, he listed five minimum conditions: "1) Meetings and exchanges must be fostered in a climate conducive to free association of, and unfettered communication between, individual scientists. They must not be under the scrutiny of secret police. 2) Participants in cooperative endeavors must be selected solely on the basis of their scientific achievements, without regard for their political conformity, race, or ethnic background. 3) Negotiations and arrangements should be in the hands of scientists, not governments. 4) Those who are invited by the host country must be allowed to accept. 5) Science areas chosen for collaboration or exchanges must offer prospects of benefit to both parties."

Other scientists were invited to give testimony to Congress on the issue of scientific exchanges, and many of these scientists expressed sentiments consistent with those expressed by Flory. It was the growing consensus in the American scientific community in the early 1980s that the "business as usual" approach of previous years in scientific relations with the Soviet Union was no longer justified or applicable. Formal scientific agreements, however, did not end. They were certainly much reduced in scope, largely because individual scientists opted out of participation. Thus, even though there were formal agreements for cooperation in place, not much actually happened in these exchange programs during the time of the scientists' boycott. The boycott was even recognized in legislation authorizing appropriations for the U.S. National Science Foundation (NSF), the U.S. government agency that provides most of the federal funding that goes to university researchers. The NSF authorization bill for fiscal year 1980, for example, made note of the SOS boycott, and requested that the NSF notify Congress before it (NSF) spends any of its appropriated money on "any scientific facilities supported by NSF and located in a foreign nation. The Committee (on Science and Technology) makes this request so that it can consider whether such construction or modification is consistent with the human rights objectives of U.S. foreign policy."[12] Thus, the scientists' boycott, even though it was a grass-roots movement completely independent of the U.S. government, did directly affect U.S. government policy.

The question remains, however, as to how truly effective the boycott was. As already noted, in the immediate aftermath of the boycott the situation for Sakharov and other dissident and refusenik scientists actually got worse, not better. Repression, in general, in the Soviet Union intensified as East-West relations deteriorated. What the boycott needed for success was for it to be sustained long enough to begin to have a noticeable adverse effect on the economic well-being of the Soviet Union, and for people in leadership positions in the Soviet government to recognize the linkage. Both of these things did indeed happen, and on a time scale much faster than anyone in either country could anticipate or predict. For, nearly simultaneous to the simultaneous SOS press conferences announcing the scientists' boycott in October 1980 another event took place in Moscow that went essentially unnoticed in the world press, an event that had a significance that almost no one then foresaw. In

October 1980 a little-known Communist Party bureaucrat named Mikhail Gorbachev was promoted to full membership in the Soviet Politburo, the "inner circle" of Soviet government leadership.

1. The experiences of Norman Zabusky, a mathematical physicist at Rutgers University, are described by Richard J. Seltzer, Washington correspondent, in *Chemical and Engineering News*, pp. 22-24, 19 March 1984. Zabusky, in the U.S.S.R. as a participant in an official U.S. - U.S.S.R. scientific exchange program, was ordered out of the country by the Soviet Academy of Sciences for conduct "inconsistent with – status as a guest" of the Academy.
2. See, for example, C. B. Afinsen, O. Chamberlain, M. Delbruck, P. J. Flory, and E. M. McMillan, *Science* Vol. 205, page 854, 31 August 1979.
3. See, for example, D. Baltimore, P. J. Flory, A. Kornberg, P. Kusch, M. W. Nirenberg, A. A. Penzias, M. Kac, and J. Cohen, *Science* Vol. 216, page 360, 1982; see also an article by R. Gillette, *Los Angeles Times*, page 10, 24 February 1982.
4. This information comes from personal communications with scientists at Lawrence Berkeley National Laboratory who did an independent analysis of the citation indices.
5. "A Report of the ACM Committee on Scientific Freedom and Human Rights," *Communications of the ACM Vol. 25(12)*, pages 888-894, December 1982.
6. "A Report of the ACM Committee on Scientific Freedom and Human Rights," *Communications of the ACM Vol. 28(1)*, pages 69-78, January 1985.
7. See report in *Physics Today*, October 1979.
8. *AMSTAT News No. 114*, April 1985.
9. From the 1978 Annual Report of the Committee on Scientific Freedom and Responsibility of the American Association for the Advancement of Science, 1515 Massachusetts Ave.

N.W., Washington, D.C. 20005.

10. The IAU incident, and the host's response, is detailed in *Physics Today*, pages 82-85, February 1980.
11. C. B. Afinsen, O. Chamberlain, M. Delbruck, P. J. Flory, and E. M. McMillan, *Science* Vol. 205, page 854, 31 August 1979; *Chemical and Engineering News*, page 32, 27 August 1979.
12. "Authorizing Appropriations to the National Science Foundation", *Congressional Record*, 96th Congress 1st Session, Report No. 96-61, March 21, 1979.

Chapter 8

Socialism with a Human Face

"I believe that to protect innocent persons it is permissible and, in many cases, necessary to adopt extraordinary measures such as interruption of scientific contacts or other types of boycotts." So wrote Andrei Sakharov from exile in Gorky in 1981.[1] He continued: "I urge the use, as well, of all the possibilities of publicity and of diplomacy. In addressing the Soviet leaders, it is important to take into account that they do not know about – and probably do not want to know about – most letters and appeals directed to them. Therefore, personal interventions by Western officials who meet with their Soviet counterparts have particular significance. Western scientists should use their influence to press for such interventions.

"Western scientists face no threat of prison or labor camp for public stands; they cannot be bribed by an offer of foreign travel to forsake such activity. But this in no way diminishes their responsibility. Some Western intellectuals warn against social involvement as a form of politics. But I am not speaking about a struggle for power. This is not politics. It is a struggle to preserve peace and those ethical values which have been developed as our civilization evolved. By their example and by their fate, prisoners of conscience affirm that the defense of individual victims of violence, the defense of mankind's lasting interests are the responsibility of every scientist."

Every Soviet dissident and refusenik scientist, without exception, supported the scientist's boycott of the Soviet Union. They were the ones who, of course, were the intended beneficiaries. Beyond these personal concerns, however, they all also understood the fundamental truths of which Sakharov wrote: the responsibility a scientist bears to the wider society in which he or she lives, because his or her special knowledge requires it. Who should understand this better than those people, like Sakharov, who were the victims of police state repression?

At the time of the announcement of the SOS boycott, Sakharov had already been in internal exile for 10 months. He and his wife, Yelena Bonnor, were confined in a very small three-room apartment in the closed central Russian industrial city of Gorky (Apartment 3, 214 Gagarin Street). "Closed" city in Soviet parlance meant that foreigners were barred from entering the city. The KGB had a squad of agents located in an apartment building about 20 yards away from Sakharov's apartment, and also in a nearby trailer. From these locations, all activity in and around the Sakharov's apartment was monitored. A jamming station was in operation to block all radio and TV reception at the apartment. There was a bath and a gas stove in the apartment, but no telephone. A police guard was posted 24 hours a day in the corridor outside the apartment to intercept and turn away all visitors. All mail was intercepted, so only a very small fraction of correspondence sent to him actually reached him, heavily censored. Every time Sakharov left the apartment to take a walk or to buy groceries KGB agents followed him everywhere, and made sure to prevent his access to public telephones and public transport, and making contact with people on the street. Of course, he had no access to libraries and no contact with other scientists. His scientific articles were forbidden to be published in Moscow or translated into English, and even his scientific papers were semi-periodically confiscated by the KGB, with agents even surreptitiously entering his apartment to rummage through his papers.[2]

What was most painful to Sakharov about this forced isolation was the effect on his health and on his family. His children, stepchildren, and grandchildren had been subjected to virtually continual KGB harassment and threats, compelling Sakharov to convince them to emigrate. "This was not a simple decision," wrote Sakharov to Anatoly Alexandrov, President of the Soviet Academy of Sciences in 1980, "and to this day is felt as a tragedy." His stepdaughter was dismissed from the school of journalism at Moscow State University, her husband was dismissed from his job, and his stepson was denied admission to the university. The children did finally manage to secure visas to emigrate. His son's fiancée, however, was not allowed to leave with him. Sakharov feared greatly for the safety of the young lady, Elizaveta Alekseeva. He pleaded with Alexandrov to help, saying that Elizaveta "has been subjected to threats and blackmail by the KGB and, a member of our family, she has not been allowed to visit me in Gorky. – In point of fact Liza Alekseeva has become a hostage."

It was intolerable to Sakharov that the KGB was abusing his family members because of the Soviet government's displeasure with him, especially since he had been exiled without benefit of any judicial proceedings whatsoever. He had simply been grabbed off the street in Moscow in January 1980, arrested, and shipped off to Gorky without charges or trial.

In 1981, without receiving any satisfactory response to his appeal to Alexandrov, Andrei Sakharov and his wife embarked on a hunger strike to protest the KGB's treatment of Elizaveta. After 16 days of worldwide outrage and condemnation, the Soviet government issued a visa to Elizaveta, and this first hunger strike came to an end. Sakharov was over 60 years old at the time, and not in great health to begin with. The next year he began to suffer from a serious heart ailment, for which treatment was not available in Gorky. He was denied permission to travel to Moscow for treatment.

Meanwhile, his wife was suffering from her own medical problems. Before the exile began, she had already traveled to Italy in 1975, 1977, and 1979 to receive treatment for glaucoma. She had served as a battlefield nurse in the Soviet Army in World War II and suffered a concussion during this time that led to loss of vision in one eye and progressive blindness in the other. She went abroad for treatment partly because the required treatment for her eye ailments was not readily available in the Soviet Union, but also because Sakharov feared for her safety if she were to be confined in a Soviet hospital. At about the same time Sakharov began having his heart problems, in 1982, his wife submitted the required forms to seek permission to go abroad for follow-up treatment for her eye problems with the same doctor in Italy who had treated her previously. While waiting for a reply she suffered the first of several heart attacks. Since she was not officially exiled, she was able to travel back and forth between Gorky and Moscow, and did receive initial diagnosis and treatment at the private medical facility of the Academy of Sciences in Moscow. During this time, however, she was the subject of an escalating campaign of attacks in the Soviet press. She was accused of being a CIA agent, a traitor, and – the most damaging charge of all – a "Zionist." She was portrayed in the Soviet press as the real radical who had led Sakharov astray. Then, in May 1984, more than a year and a half since first applying for a visa to go abroad for medical treatment, she went to the airport in Gorky, accompanied by Sakharov, to get a flight to Moscow, after being promised an answer to her visa application. While Sakharov watched from the airport window, his wife was grabbed by KGB agents as she approached the aircraft and whisked away. Sakharov went back to his apartment and immediately began his second hunger strike. While his wife was undergoing frequent interrogations by the KGB, Sakharov was seized by KGB agents and taken to Gorky Regional hospital, where he underwent a four-month ordeal of forced feedings and general maltreatment. A few days after his forced confinement in the Gorky hospital he suffered his first stroke.

Meanwhile, his wife, unbeknownst to him, was tried and convicted of anti-Soviet slander and sentenced to five years of internal exile – in Gorky. So, while she was held under house arrest at the Gorky apartment, unable to receive the medical care she required, Sakharov was held in virtual isolation at the nearby hospital, unaware of where his wife was or what had happened to her. All of this took place in the summer of 1984. After he ended his hunger strike and returned to the apartment, approximately four months after it first began, he wrote another long letter to Alexandrov, detailing his experiences of the previous months and asking once again for Alexandrov's help. He noted in his letter the "remarkable coincidence" of his experiences with those of the protagonist in George Orwell's novel *1984*, about life in a totalitarian state of the future.

During Sakharov's ordeal in the summer of 1984, not much reliable information reached the West about his condition. The Soviet authorities, of course, used his isolation as an opportunity to spin their own propaganda, again portraying Sakharov as a demented and misguided individual, who – far from being maltreated – was in fact receiving the care he needed. Almost as soon as the hunger strike began in May 1984 the Soviet news agency *Tass* denied it. Then in August videotapes were obtained that showed Sakharov eating and in good health and good spirits. It became clear, however, that these videos, which were purported to be current ones, were in fact crude forgeries by the KGB, made by splicing together still photos and earlier videos. The effect of the official Soviet portrayals was to increase the alarm in the West over Sakharov's fate. It was already well known that psychiatric treatments in the Soviet Union had been subverted as an instrument of political repression of dissidents. By 1984 some 200 dissidents had been confined to Soviet psychiatric hospitals, where they were often treated to potent psychotropic drugs. Indeed, just the previous year the U.S.S.R. withdrew its membership in the World Psychiatric Association in the face of an effort that was bound to succeed to expel them.

Given this history, Western scientists were fearful that Sakharov was being subjected to psychiatric drugs.[3] Of course, there was no way at the time to know whether or not these fears had any foundation in fact.

While Sakharov was enduring these horrors in 1984 there was another event that took place in the United States that would ultimately have a bearing on his fate, though no one could foresee this at the time. This event was the invention and introduction of the Macintosh personal computer.

We remember from the last chapter that the work of de Broglie in France and Bohr in Denmark led to the idea that fundamental particles like electrons behave on the microscopic scale like waves. The subsequent development of "wave mechanics" in the early part of the twentieth century by many scientists worldwide was successful in explaining the fundamental structure of atoms. Wave mechanics is actually a mathematical formalism – several formalisms, in fact – that provide scientists with a prescription for how to put together computational models of atomic systems. These "quantum" models proved to be valid when their predictions could be verified by experiment.

One of the many great successes of wave mechanics was in providing an explanation for why some chemical elements are metals with a high electrical conductivity, some are insulators with very little conductivity, and others fall in between and are thus called semiconductors. It all has to do with how the atoms of the material, held together by electrical forces, arrange themselves in a lattice structure according to the laws of wave mechanics, and how the electrons in the outer orbits of these atoms (the so-called valence or conduction electrons) diffract like waves in this lattice. This diffraction results in the valence electrons being in only certain allowed energy bands. Depending on where these energy bands are with respect to the energy at which they are essentially unbound from the parent atom and thus can travel more or less freely through the lattice, the material is a conducting metal, a semiconductor, or an insulator.

This new understanding of material structure gave birth to whole new industries in the developed Western countries. Scientific research became the engine of technological innovation and development, which in turn became the engine of the economy and its modernization. Major industries established their own corporate research laboratories, some of them devoted to basic research as well as to technology and product development. Corporate laboratories like the Bell Telephone Laboratories and the General Electric laboratory became well-known leaders in scientific research. It was Bell Labs researchers who, exploring the properties of semiconductors, developed a device they called a transistor. This small wafer of semiconductor material was able to behave much like the vacuum tubes in electronic devices like radios, but required much less electrical power, much less heat dissipation, and was substantially smaller.

The development of the transistor revolutionized the electronics industry, and made possible numerous future technological developments. Arguably the most important of these was the miniaturization of the electronic computer.

The idea of a personal computer originated with Steve Jobs, a college dropout who had worked first at the Hewlett-Packard (HP) electronics company and then at the Atari video game company. Jobs teamed up with another former HP employee, Steven Wozniak, to put together the first personal computer, the Apple I, in the garage of Jobs' house in California, taking advantage of the new technology for making computer memory circuits out of small semiconductor transistors. It was the first single-board computer with a built-in video interface and an internal memory that told the machine how to load other programs from an external source. The Apple I was first marketed in 1976. By 1983 when IBM introduced to the market their own version of a personal computer, the Apple Computer Company had experienced a compound growth rate of 150% per year. The Macintosh, introduced in 1984, had more memory, a faster microprocessor, and a mouse-driven, user-friendly interface.

The development and marketing of the personal computer were very significant for several reasons. It was, first of all, a clear demonstration of the advantages of a free and open system that not only allowed ordinary citizens access to information and technology, but actually encouraged exploration and innovation. Not only could everyone freely access information, they now had, with the personal computer, the opportunity to create their own information. More significant for this story, though, is the fact that the personal computer became a serious threat to the viability of a police state, like the Soviet Union, and its central control of information. In a country where even the personal ownership of a typewriter was considered a subversive act, the personal computer, and what it represented in terms of personal freedom, was an alarming development. This was especially true as it became clear – and it did become clear quite quickly to a few reform-minded leaders in the Soviet Union – that the police state was incompatible with the technological innovation on which the modernization of the economy depended. Without modernization of the economy, there was no way to sustain the state's power. This was a real dilemma for the Soviet Union. It was a dilemma that few within the Soviet Union recognized, and even fewer were prepared to deal with it.

Leonid Brezhnev was incapable of understanding the problem his country faced. When he died in 1982 after 17 years as General Secretary of the Communist Party only Stalin had served in the top leadership position of the U.S.S.R. for a longer period of time (1923 - 1953). There was little hope for reform when the aged chief of the KGB, Yuri Andropov, was chosen by the Politburo to succeed Brezhnev. When Andropov died just fifteen months after taking office, the Politburo, instead of choosing Andropov's protégé Mikhail Gorbachev, elevated another aging hardline conservative, Konstantin Chernenko, to the top post. It was during this time period that Sakharov underwent his ordeals in Gorky, and, in general, East-West relations further deteriorated while repression of dissidents – particularly dissident scientists – intensified.

Chernenko, like Andropov before him, did not last very long. At his death in 1985 the Politburo then selected Mikhail Gorbachev as the new General Secretary. Even though Gorbachev's reformist tendencies were already in evidence by 1985, it has been argued persuasively that the old oligarchs who made up the membership of the Politburo at the time were really clueless about the consequences of what they had just done[4].

Gorbachev, born on March 2, 1931, was just ten years younger than Sakharov. He came from a family of peasant farmers in the Stavropol region of Russia, where he first joined the Communist Party at age 21. He received his law degree from Moscow State University in 1955 and then returned to Stavropol, where he moved up the ranks in the regional Communist Party hierarchy. He first served as First Secretary of the Stavropol Young Communist League, then First Secretary of the Stavropol City Communist Party Committee, then First Secretary of the Stavropol Regional Communist Party Committee. During these early years, he completed a course of study at the Stavropol Agricultural Institute, and issues relating to the agricultural economy of the Soviet Union became his specialty. Thus, seven years after his elevation to membership in the Central Committee of the Communist Party of the Soviet Union, in 1978, he was chosen to fill the newly vacant post of Agriculture Secretary.

Perhaps this is why his selection as General Secretary in 1985 came as a surprise to so many people. The agriculture post was viewed mostly as a dead-end career path in the Communist Party hierarchy. Ever since the massive upheavals and mass murders of Stalin's forced collectivization of agriculture, food production in the Soviet Union was notoriously inefficient, and food distribution even worse. The Soviet Union's inability to feed its own people adequately while passing itself off as the world's model of socialist triumph was the subject of endless jokes. One such joke:

Two Russian friends meet on the street. One says to the other, "Did you hear the news? Brezhnev has just won the Nobel Prize for Agriculture!"

"Really?" asks his friend. "What did he do?"

The first friend replies: "He planted wheat in Kazakhstan and reaped the harvest in Kansas."

Thus, when something went wrong with agricultural production in the Soviet Union, as it invariably did, it was not unusual for the Agriculture Secretary to take the blame. The Agriculture Secretary post was the place the Central Committee sent one of its members in order to derail his career. Indeed, every year after Gorbachev took over the Agriculture post was a bad harvest year in the Soviet Union, with food imports increasing. So, it is not surprising that Gorbachev's selection to succeed Chernenko as General Secretary would come as a surprise. It *was* a surprise. It was a surprise not only to American foreign policy experts who had not anticipated it, but even to the Soviet Politburo, for even though the signs were present before Gorbachev's selection of his reformist tendencies it is unlikely that the entrenched conservatives in the Politburo had any clear idea of what they had just unleashed.

Gorbachev set himself the task of restructuring and modernizing the Soviet economy. He had no intention of dismantling Communism, but he did recognize that one of the main obstacles to the reforms he proposed was the corruption of the Communist Party's entrenched bureaucracy. Even before becoming General Secretary Gorbachev had set himself the task of making personnel changes in Party and government offices, supervising regional elections to bring in younger and more able officials in areas where economic performance was poor or where corruption was thriving. He had also worked to change the management of the agricultural economy by putting food production, food processing, and food distribution under common management. The Food Programs he devised were designed to reverse the fortunes of Soviet agriculture, but after another disastrous harvest in 1982 Gorbachev's political future looked very bleak. It is likely that the only thing that saved him from political oblivion at this point was the death of Leonid Brezhnev on November 10 of that year. What followed was a two-and-a-half-year period of nearly complete dysfunction of the Soviet government, of which mismanagement of agriculture was only one component.

Gorbachev, however, understood clearly the linkages between reform and democratization. Indeed, as early as a Party conference in Moscow on December 10, 1984, he made the claim that "*glasnost* (openness) is a compulsory condition of socialist democracy and a norm for public life." He was advancing the view that progress could be made only by allowing people the freedom to voice their ideas and to criticize government, a view remarkably similar to Sakharov's. *Glasnost* became official policy after Gorbachev became General Secretary. In practice this meant that some press restrictions were eased, allowing people to publish letters in the newspaper, for example, criticizing poor performance or corruption of government bureaucrats. Such openness, of course, did not extend to tolerating criticism of Soviet government policy, particularly foreign policy, but this new openness was, nonetheless, a significant and important advance for what had been, until this time, a thoroughly closed and tightly controlled society.

Gorbachev also understood the importance of science and

technology to modernizing the economy and to the success of the reforms he was trying to institute. Only three months after becoming General Secretary he increased by up to 50% the salaries of research scientists, technologists, and engineers. He also instituted a system of bonus pay to reward productivity and efficiency. A similar system of bonus pay was adapted to other sectors of the economy, with salaries linked to quality of work. In November 1985 he announced a completely new reorganization of agricultural management. The following year a new policy of *Perestroika* (Restructuring) began. With this new policy the linkage between science, democratization, and economic progress was made explicit. At the Communist Party Plenum of January 1987, *Perestroika* was defined as a strategy "to unite the achievements of the scientific-technical revolution with a planned economy and to bring into action the entire potential of socialism."

"*Perestroika*" according to the Party statement, "is the buttress for the vital creativity of the masses; it is the all-sided development of democracy and socialist self-direction, the encouragement of initiative and independence, the strengthening of discipline and order, the widening of *glasnost*, criticism and self-criticism in all spheres of social life; it is a greatly heightened respect for the value and worth of the individual."

Gorbachev was, as he himself put it, trying nothing less than to create "socialism with a human face." Although this formulation sounds, on its face, much like Sakharov's, there were important differences. Gorbachev was a committed Marxist. The same Party statement on *Perestroika* that made clear that the reform movement was "to bring into action the entire potential of socialism" also defined *Perestroika* as "the re-establishment and development of the Leninist principles of democratic centralism in the direction of the national economy, – (and) the readiness and fervent desire of scientists to actively support the course of the Party for renewing society." In other words, scientists were to be recruited to reform the system but not to question its basic foundations. Freedom of expression was to go only so far.

Accordingly, Gorbachev did not see fit, in the beginning, to release Sakharov from internal exile. In the months after Gorbachev's assumption of the top leadership position in the Soviet government, he stayed close to the official party line that Sakharov, if released, presented a danger to the Soviet Union. This danger arose from Sakharov's alleged knowledge of state secrets, supposedly from his work on Soviet nuclear weapons, and the implied assumption that he would leak these secrets to foreigners if he were to be released. It seems not to have occurred even to Gorbachev that Sakharov had not worked on nuclear weapons design for more than 20 years by this time, so certainly anything he knew was technically obsolete.

Gorbachev wanted to reform the system, not replace it. He was, after all, still a creature of the system. Thus, even though he came into the leadership position with ideas for solving the Soviet Union's "nationalities question," little or no real progress was made on this front. He was still stuck with old Party attitudes. Jewish emigration was largely restricted just as before. Repression of ethnic minorities in the fifteen republics of the Soviet Union continued much like it did before, and in some instances even intensified. Ethnic unrest and pro-independence demonstrations bubbled up in the Baltic republics, in Georgia, and even in Ukraine. Gorbachev opposed independence for the republics, but, to his credit, he did not send in the tanks as Brezhnev had done in Czechoslovakia in 1968.

Gorbachev recognized that the maintenance of such a large military occupation of Eastern Europe was an unacceptable drain on the country's economic well-being. He also believed in the abolition of nuclear weapons. In a visit to Prague in April 1987 he called for the elimination of all nuclear weapons from Europe. Before he could get to this point, which led to breakthroughs in nuclear arms control with the United States, several things happened first to set him on the course that ended the Cold War, and Andrei Sakharov played the key role.

For nearly two years after Gorbachev ascended to power in the Soviet Union, Sakharov continued in his isolation in his Gorky apartment. Sakharov was barred from laboratories and libraries, so he found it painfully difficult to continue to function as a scientist. Despite this, his scientific stature rose in the West, largely because the attention generated by his plight also brought attention to his scientific work, and Western scientists began to gain a greater appreciation for the depth and breadth of his scientific thinking. Scientific papers that he could not get published in the Soviet Union found their way, by a variety of routes, to the United States, where a number of scientists and scientific institutions – most notably Stanford Linear Accelerator Center in California – set about translating them and getting them published. Some postcards that Sakharov wrote in response to reprints of scientific articles he received that managed to get past the censors also brought attention to his thinking on a couple of the hottest scientific questions of the day.

Sakharov was then focusing his scientific attention on fundamental questions about nuclear structure and the nature of elementary particles. As discussed earlier, scientists had already developed an understanding of electric and magnetic forces, and shown that light, radio waves, heat waves, and x-rays are all different manifestations of the same fundamental thing: electromagnetic waves, or photons, that mediate the electric force between electrically charged particles. Quantum theory had extended electromagnetic theory to show that the atomic structure of all matter could be explained by considering electrons as elementary charged particles, with unit negative electric charge, that behave like waves. It is also the electric force that holds electrons in their orbits inside atoms, and the electric force that hold atoms together inside solid materials. Quantum theory was a spectacular success in explaining the individual structure of individual atoms, and the lattice structure of individual atomic arrangements within solids. Quantum theory is able to tell us why the particular chemical elements and materials we see in nature do exist and why other arrangements that we do not see in nature cannot exist. Scientists, however, have not had as complete a picture of how the nucleus of atoms is held together.

All atomic nuclei are composed of protons and neutrons. Both of these particles are about 2000 times more massive than an electron. The proton has a unit positive charge, and the neutron is uncharged. Thus, a neutral atom of any given element contains a particular number of protons, with an equal number of electrons in fixed orbits around the nucleus. All carbon atoms, for example, have six protons in their nuclei, and, of course, six electrons orbit the nucleus in the neutral atom. The carbon-12 atom, which comprises most of the carbon in nature, has six neutrons in addition to the six protons in its nucleus, so its atomic weight is 12. Carbon-13, which also has six protons in its nucleus, has seven neutrons. These two forms of carbon are *isotopes*, which means they are just different forms of the same element, containing the same number of protons but a different number of neutrons.

Most of the mass of the atom, then, is in the nucleus, but the size of the atom is fixed by the size of the electron orbits, so all atoms are mostly "empty" space. The electrons are bound to the nucleus by electric forces, the attraction of the positive charge of the protons for the negative charge of the electrons. The protons and neutrons, however, are bound together in the nucleus by a much stronger nuclear force. The nuclear force, unlike the electric force, acts only over a very short range. To come within the range of a nuclear particle so as to feel the effects of the nuclear force requires scientists to accelerate nuclear particles like protons and neutrons to very high velocities, very high energies, and aim them at other particles. When such particles smash into each other and interact with each other via the nuclear force we observe many other "elementary" particles that are created. A fundamental question that Sakharov and other scientists wrestled with was whether this new "zoo" of particles was indeed elementary and their existence and properties could be explained by some new "field" theory – analogous to the explanation of the existence and behavior of atoms by electromagnetic quantum field theory – or if each of the particles in the zoo, including the proton and neutron, were all built up by even more fundamental elementary particles that we have not yet observed. Sakharov favored the latter hypothesis. He had written a number of early papers that proposed explanations for the results of particular accelerator experiments on the picture that protons and neutrons, as well as all the other "elementary" particles, are really composed of different combinations of three elementary "quarks". Some of Sakharov's early publications provided support for the quark theory. This work was just becoming known in the West during the time of his exile in Gorky, also a time when new experimental results at a number of different laboratories were beginning to lend added credence to the quark theory. Thus, this work of Sakharov helped considerably to solidify his reputation as a first-rate scientist. If the general public knew of Sakharov only as a human rights activist and, perhaps, as inventor of the Soviet nuclear bomb, to the Western scientific community he was this and much more.

Sakharov's work on the quark theory led to his proposed solution of what had been one of the most fundamental puzzles about the structure of the universe. It had been known for many years, and is a confirmed prediction of quantum theory, that for all elementary charged particles that exist there also can exist an anti-particle that has the same mass but the opposite electric charge. According to quantum theory, electrons with unit negative charge can exist – and indeed we observe them readily in nature – but "positrons" can also exist, particles that are identical to electrons in every way except in having a unit positive, instead of negative, charge. Yet, we do not see positrons in nature. We can only make them in accelerator experiments, and then they do not last very long, because whenever they "bump" into electrons the two oppositely charged particles annihilate one another, creating a high-energy photon. The same is true for anti-protons. If everything we observe in the universe – stars, galaxies, planets, rocks, people – is made of matter, and there is no anti-matter anywhere, this means that the universe started out with a basic asymmetry. It had to contain more matter than anti-matter in the beginning. But basic quantum theory tells us that the primordial "soup" of elementary particles from which the universe was created could only have equal numbers of particles and anti-particles.

Sakharov proposed a solution to this conundrum. He showed that if the proton is unstable, meaning that it can disintegrate into other particles after a certain time, one can explain the matter/anti-matter asymmetry. He figured that the proton would not have to be very unstable in order to explain the current universe, so there is no danger any time soon of atoms disintegrating. Sakharov's hypothesis was an extremely important advance. It led to new experimental work to look for evidence of the predicted proton decay, and gave birth to a whole new field of scientific investigation with the linking of nuclear and particle physics with cosmology.

It was as much because of Sakharov's importance to science as his importance to the cause of scientific freedom that the scientists' boycott of the Soviet Union continued and intensified as long as Sakharov was confined in internal exile. Even with all Gorbachev's reforms, which were certainly viewed in the scientific community as a positive development, the boycott was still on.

Meanwhile, it must have become clearer to Gorbachev that the boycott was not going to be forgotten or abandoned anytime soon. He was, at the same time, becoming less tolerant of the boycott, because it was interfering with the business of reform and modernizing of the Soviet economy. Gorbachev needed certain things from the West, things he was not getting because of the boycott. He needed high-technology imports, like computers, high-precision machining tools, electronics of all kinds. He needed access to scientific and technological information. In order for Soviet scientists to play the critical role in *Perestroika* that he had defined for them, he needed for them to be fully integrated into the world community of scientists. None of this was happening. For all of these reasons, he needed to find a way to end the boycott.

Most of all, he needed to overcome the stalemate in arms control and find a way to reverse the Cold War's seemingly endless nuclear arms race. Continuation of the Cold War was a major obstacle to *Perestroika*. To break through this stalemate, he needed Andrei Sakharov. It was only a matter of time before he would realize this.

Fortunately, it did not take him too long to come to this realization. Perhaps his decision was accelerated by the nuclear reactor meltdown and explosion that occurred at Chernobyl in April 1986, when the Soviet Union's need for Western technological help became apparent to all the world[5]. Perhaps he saw some merit in the arguments Sakharov made in two letters the scientist sent to Gorbachev, one in February 1986 and one in October. In the second letter, Sakharov told Gorbachev that he had been detained and exiled illegally. For whatever reason, Gorbachev reversed his original position on Sakharov and freed him from exile in December 1986, only about a year and a half after first coming to power, and only a few months into *Perestroika*. In fact, Gorbachev spoke directly to Sakharov by telephone on December 16, informing him of his decision. Once freed, Sakharov returned immediately to Moscow, and the pace of events leading to the end of the Cold War greatly accelerated.

Just weeks after Sakharov's release from exile he spoke at a three-day international disarmament conference held in Moscow in mid-February 1987. In this speech, he emphasized the same points he had made earlier concerning the American development of a ballistic missile defense, the so-called Strategic Defense Initiative. According to Sakharov, SDI was a costly waste of money that was not likely to be successful, and that, in any case, could be easily and cheaply defeated by decoys and other offensive strategies of the Soviet Union. Therefore, it made no sense for progress in nuclear arms control to be held hostage to Soviet demands that the Americans end their SDI program. Let the Americans waste their money, said Sakharov; the Soviet Union should de-link SDI from arms control. This same point was made by two American scientists, Frank von Hippel and Jeremy Stone, who also attended the conference.

Present and listening to this speech was Mikhail Gorbachev. The Soviet leader later wrote in his *Memoirs*[6] that this was his first face-to-face meeting with Sakharov, while acknowledging that he had talked to the scientist on the telephone earlier. He also confirmed in his *Memoirs* that he viewed Soviet scientists, and Sakharov in particular, as essential to his *perestroika* reform program, although he makes no mention of why specifically he decided to release Sakharov from his internal exile. He also makes no mention of the scientists' boycott.

Sakharov's proposal for de-linking SDI and arms control went counter to then-current Soviet policy. Gorbachev, however, listened. He also listened intently to the two American scientists, von Hippel and Stone. He reportedly said later that his discussions with the scientists at the disarmament conference "made a big impression" on him.[7] As a result, he then adopted Sakharov's proposal as his own, which he discussed at a Politburo meeting shortly after the conclusion of the disarmament conference. Then, on February 28, 1987, the Soviet government announced that it was prepared to reach an agreement with the United States on a proposal to reduce or eliminate intermediate-range nuclear missiles. Just two months later, in April 1987, Gorbachev called for the elimination of all nuclear weapons from Europe in a speech while visiting Prague, Czechoslovakia. This was just the breakthrough that was needed in moving arms control negotiations forward.

The first focus of the effort to eliminate nuclear weapons came with the negotiation of the Intermediate Nuclear Forces (INF) Treaty. Such nuclear forces had been a major sticking point in arms control negotiations throughout the Cold War. Soviet development of intercontinental ballistic missiles capable of threatening the United States with nuclear weapons had grown enormously during the 1960s. By 1966 the Soviet Union had also begun to deploy an anti-ballistic missile defensive system around Moscow, a development that triggered a further escalation in the nuclear arms race. Complicating the picture for American defense planners was the successful test of a nuclear missile in 1966 by the People's Republic of China. In the 1970s the Soviet Union achieved approximate nuclear weapons parity with the United States, this at a time when ongoing negotiations on treaties to limit nuclear weapons development and deployment continued to be stalled over several issues, a major one of which was the Soviet position on how intermediate-range nuclear forces were to be counted in any agreement. On the very day that Richard Nixon was inaugurated as President of the United States, January 20, 1969, the Soviet Foreign Ministry sent a message to the United States signaling Soviet willingness to discuss limitations on strategic nuclear weapons systems. The problem, though, was that the Soviets defined "strategic" as any weapon that could reach the territory of the other country, so the Soviet Union wished to count all "forward-based" U.S. nuclear weapons systems, which meant to them short-range and intermediate-range American bombers based on aircraft carriers or in Europe, but not count Soviet intermediate-range missiles that targeted Western Europe. In the 1970s the Soviet Union began to replace their older SS-4 and SS-5 intermediate-range nuclear missiles with the more-modern SS-20. Unlike the older missiles, the SS-20 was capable of carrying three independently targeted nuclear warheads. It was also capable of being carried on a mobile launcher, was much more accurate, and could be easily concealed and rapidly re-deployed. The United States saw the SS-20 deployment as a significant change in the European security situation, and a threat to the NATO military position.

The American and NATO response to Soviet deployment of their new SS-20 intermediate-range ballistic missiles in the western part of the Soviet Union was a corresponding deployment of American intermediate-range missiles in Europe. This deployment was part of a "dual track" strategy. The first track continued negotiations to limit intermediate-range forces on both sides, and the second track called for deployment of 464 single-warhead ground-launched cruise missiles (GLCMs) and 108 Pershing II ballistic missiles in Europe, beginning in 1983. The American deployment had been resisted by large segments of the European public and European governments, but much of this opposition either evaporated or was overcome after the Soviet invasion of Afghanistan, and the increased repression and intensified Cold War confrontations that followed in its wake. While deployment proceeded, President Reagan announced on November 18, 1981, a new proposal – consistent with the first track of the "dual-track" strategy – to eliminate the entire fleet of GLCM's and Pershing II missiles in exchange for the Soviet Union eliminating all their SS-4's, SS-

5's, and SS-20's. This proposal became known as the "zero-zero" offer. On-again, off-again negotiations on this proposal and its numerous variants did not come to fruition until those eventful weeks after Sakharov's speech, and Gorbachev's response, in the first months of 1987. In the course of meetings with U.S. Secretary of State George Schultz in Moscow in April 1987 Gorbachev revived the "zero-zero" option. In July he agreed to accept "zero-zero," and Soviet negotiators were instructed to work out the details of an agreement based on this proposal. The Treaty was completed and signed by President Reagan and General Secretary Gorbachev on December 8 during the course of a summit meeting in Washington, D.C. The INF Treaty called for the removal from Europe of all missiles carrying nuclear warheads with a range of between 500 and 5500 kilometers. By the Spring of 1991, as called for by the INF Treaty, all such missiles were dismantled, 2692 missiles in all, a major breakthrough in ending the Cold War.

The next focus was on strategic nuclear weapons, i.e., nuclear weapons carried on intercontinental ballistic missiles, on long-range bombers, and on submarines. These strategic weapons are the backbone of the policy of deterrence, for they allow each side to threaten the destruction of the other by launching weapons from their own territory or from the open sea at a great distance. Gorbachev wanted not only to eliminate intermediate-range nuclear weapons from Europe, but intercontinental-range ones based in America and Russia as well. The mechanism for advancing toward this goal became the Strategic Arms Reduction Treaty (START) negotiations. The goal of START was to reduce the numbers of strategic nuclear weapons on both sides in a phased and verified manner. Nuclear weapons reduction meant more than just removing the weapons from the missiles, bombers, and submarines. It meant actual dismantlement of the weapons, a matter to be discussed in more detail in the next chapter.

The idea for START was quite different than the ideas for a nuclear weapons freeze pushed by the bilateralists, as discussed in Chapter 6. The bilateralists pushed for the U.S. to ignore Soviet human rights abuses in order to advance their self-contradictory arms control positions, not recognizing that these two things are, and should be, integrally linked. START, on the other hand, was independent of bilateralism because it did not tie the interests of either side to unrealistic assumptions about the other. In this sense, it was similar to, and an outgrowth of, the build-down proposal, as first advanced by Senators Sam Nunn (a Democrat from Georgia) and William Cohen (a Republican from Maine, who later became Secretary of Defense in the Clinton Administration) and discussed in detail by Alton Frye in a journal article,[8] and, in a different context, by the American scientist Carl Sagan.[9] In the build-down proposal, for every new, modern nuclear weapon added to the U.S. stockpile two or more older weapons were to be retired and dismantled. Such a policy was meant to allow the modernization of the U.S. strategic forces, maintaining the credibility of the strategic deterrent, while at the same time decreasing the numbers and total destructive power of the U.S. nuclear arsenal. Most important, it was meant to be a policy that could be pursued independently of whatever steps the Soviet Union chose to take or not take toward control of its own nuclear weapons. This is because, unlike previous arms control initiatives, the build-down proposal did not seek a fundamental restructuring of Soviet forces as a precondition to any U.S. moves.

The build-down concept was endorsed by President Reagan in a speech on October 4, 1983. Critics of the Administration, however, saw this endorsement as a cynical and not so subtle attempt to enlist Congressional support for the President's plan for MX missile deployment. Indeed, the build-down proposal was worked into a historic compromise between Congress and the President by the Scowcroft Commission. The compromise allowed the President to get Congressional funding approval for the MX in exchange for assurances that he would propose bold new arms control initiatives to the Soviets that would incorporate build-down. The President did not, at that time, fulfill his end of the bargain, which confirmed Administration critics in their view that President Reagan saw arms control as intrinsically inimical to U.S. interests and instead sought to pursue a policy of achieving military dominance at any cost. Thus, the ideological rigidity of the Reagan Administration, coupled with the ideological rigidity of the Soviets, contributed to the initial breakdown of the INF and START talks in the pre-Gorbachev years.

All of this changed after Gorbachev adopted Sakharov's proposal for de-linking SDI from arms control. Soviet ideological rigidity was no longer an obstacle to progress in arms control, because ideological rigidity was being replaced in the Soviet Union, at long last, by Sakharov's rationalism. Further, President Reagan was receptive to the new Soviet flexibility, a new position that perhaps surprised and confounded both his supporters and his critics. It is likely, though, that the President genuinely did share Gorbachev's dream of ridding the world of nuclear weapons.

After this initial breakthrough, events leading to a stand-down of the Cold War confrontation followed in quick succession. Gorbachev announced the beginning of Soviet troop withdrawal from Afghanistan in January 1988 (another action urged by Sakharov), a process that was completed the following year. In May 1988 President Reagan met with Gorbachev in Moscow. START I was signed in Moscow on July 31, 1991, by Gorbachev and President George H. W. Bush. This historic arms control agreement was the first one to apply a variant of the build-down concept to strategic nuclear weapons.[10] It called for the phased reduction of the nuclear arsenals on both sides by 30%-40%, with a reduction of 50% in some of the most threatening strategic systems. The Treaty called for the actual elimination of as many as 9000 nuclear warheads on both sides, and the elimination of thousands of missiles and other strategic delivery systems as well.

Meanwhile, the Soviet Union did not, unlike 1956 in Hungary and 1968 in Czechoslovakia, send in the Soviet army as, one by one, the Warsaw pact countries of Eastern Europe overthrew or voted out their Communist governments and withdrew from the Soviet political and military orbit. The Berlin Wall came down in November 1989, sparking a rapid process of Soviet troop withdrawals from all of Eastern Europe. The *status quo* posture of the George H. W. Bush Administration certainly helped; the President signaled Gorbachev in several different ways[11] that the U.S. would make no moves to inflame the anti-reform hardliners in the Soviet government and Communist Party leadership by taking advantage of Soviet troop withdrawals, giving much-needed breathing space for Gorbachev to effect his reforms in the face of serious internal opposition. Finally, an attempted Communist coup against Gorbachev in August 1991, during which coup plotters actually kidnapped him, led to the final failure of the old Communist regime and the final dissolution of the Soviet Union four months later, on Christmas Day 1991.

This remarkable chain of events was set in motion *not* because of any American "get tough" policy, *not* because American technology or political ideology was superior to that of the Russians. The Russians did not capitulate on arms control because of fear of SDI, as the American right continues to maintain. The Soviets agreed to arms control because they were *not* afraid of SDI, and because they saw these arms control agreements in their own best interest. They came to see these agreements, and the necessary winding down of the Cold War, in much the same way that Sakharov had seen them. Gorbachev came to accept the concept of linkage. His freeing of Sakharov and acceptance of Sakharov's basic theses on linkage was the trigger for all the events that followed. And it was the scientists' boycott, and its effects on the Soviet Union, that helped to provide the necessary extra "push" that Gorbachev needed to propel his thinking in the right direction.

Was the extra push provided by the scientists' boycott really necessary? Would not the Soviet Union have come undone simply as a result of the *perestroika* and *glasnost* reforms that Gorbachev instituted, as claimed by some authors?[12, 13]

It is always a dangerous undertaking to engage in counterfactual history, so I will not even speculate here as to what would have happened in the absence of the scientists' boycott. What actually happened, though, is significant, because Gorbachev's ending of the scientists' boycott by freeing Sakharov opened the way to substantially reducing the threat of nuclear war between the U.S. and the U.S.S.R., which, I claim, was a necessary pre-condition for allowing a peaceful dissolution of the Soviet Union. It is not a given that a peaceful end to the Cold War would have happened anyway if there had been no scientists' boycott and had Gorbachev not had the discussions with the scientists at the disarmament conference in February 1987.

The story does not end here, however. The Cold War may be formally ended with the end of the Soviet Union. Many things needed to happen, though, in the aftermath of the end of the Soviet Union to bring an end to the apocalyptic threat of nuclear war.

First, a highly militarized Russia, with an economy that was in the early 1990s still tied closely to military production and centralized command, needed to continue in its process of conversion to a modern market economy. Nuclear weapons withdrawn from the Russian nuclear weapons stockpile under the terms of START I needed to be dismantled. Dangerous nuclear materials from these weapons needed to be adequately safeguarded from falling into the hands of terrorists or thieves who would sell them to other nations. All this had to be done while helping the new Russian Federation stay on the right course in its political, economic, and social transformations. The next chapter discusses this first phase of the stand-down from the Cold War, and the key role that American scientists played in all of these efforts. Fortunately, further scientific developments initially aided this process, and increased confidence in its ultimate success.

Unfortunately, other parallel developments in both countries worked to frustrate these efforts, leading to a slow unraveling of the arms control regime that had so carefully been crafted over the decades of the Cold War in order to protect the world from the apocalypse of nuclear war. In addition, other apocalyptic threats to human existence arose during this same time period. These developments are discussed in Chapter 10.

Finally, in the last chapter, I summarize what lessons we have failed to learn from the end of the Cold War, lessons which we must learn if we are to have any real hope of managing the multiple apocalyptic threats facing humanity today, as we begin the third decade of the twenty-first century.

1. "The Social Responsibility of Scientists," by Andrei Sakharov, *Physics Today*, June 1981.
2. This information has been reported in numerous places, including by Sakharov himself in a letter published in *Nature Vol. 288*, page 112 (1980).
3. *Chemical and Engineering News*, July 23, 1984.
4. *The Gorbachev Factor*, by Archie Brown (Oxford University Press, 1996).
5. See, for example, *Midnight in Chernobyl: The Untold Story of the World's Greatest Nuclear Disaster* by Adam Higginbotham, Simon and Schuster, 2019.
6. *Memoirs* by Mikhail Gorbachev, Doubleday, New York,

New York, 1996.

7. *The Myth of Triumphalism: Rethinking President Reagan's Cold War Legacy* by Beth A. Fisher, University Press of Kentucky, 2020.
8. "Strategic Build-Down," by Alton Frye, *Foreign Affairs Vol. 62/2*, Winter 1983/84.
9. "Nuclear War and Climatic Catastrophe," by Carl Sagan, *Foreign Affairs Vol. 62/2*, Winter 1983/84.
10. This is not to say that there were no strategic arms control agreements between the Limited Test Ban Treaty of 1963 and START I in 1991. Every American administration during those nearly three decades, Democrat and Republican, engaged with the Soviet Union in arms control negotiations, and two Strategic Arms Limitation Treaties (SALT) were signed. The SALT treaties, however, only prohibited each side from further expanding their arsenals. It was only with the START process that both sides agreed to actual reductions.
11. *At The Highest Levels: The Inside Story of the End of the Cold War* by Michael Beschloss and Strobe Talbott, Little Brown and Company, New York, 1994
12. *The Gorbachev Phenomenon* by Moshe Lewin, University of California Press,1991. This book is largely a sociological analysis of *Glasnost*, Gorbachev's policy of "openness", one of the reforms that Gorbachev instituted that contributed to the loosening of Soviet police state controls.
13. *Dismantling Utopia* by Scott Shane, Ivan R. Dee Publishing, 1995. This book discusses how information in the form of electronic media, videos, and films that flooded into the Soviet Union after Gorbachev relaxed police state controls led to the dismantling of the Soviet Union.

Chapter 9

The Aftermath, Part I: Dismantlement and Stewardship

As I have argued, the Cold War ended in essence when Mikhail Gorbachev released Andrei Sakharov from internal exile and then accepted his basic arguments de-linking Soviet opposition to SDI from progress on arms control, and linking lifting of police state controls with economic modernization. With the lifting of police state controls the Soviet state could no longer maintain itself.

Securing a permanent cessation of the Cold War, however, was only just beginning. The challenge was then – and still is – to dismantle a highly militarized society and then reconstruct a new democratic society in its place. This is a challenge in which scientists, Russian and American, continue to play lead roles.

At the time Boris Yeltsin became the first elected President of the new Russian Federation in July 1991[1] there was no Russian equivalent of the Bell Telephone Laboratories or IBM, no commercial sector that could translate scientific advances into technologies that could impact economic growth. The Soviet command economy was organized largely for military production. Entire communities all across the eleven time zones of Russia and the other states of the Former Soviet Union (FSU) existed predominantly on the state payroll of a single industry geared to some specialized sector of military production. In these smaller communities, one would often find the large industrial plant located on one side of the main highway, with the other side occupied by a collection of tall Soviet-style cinder-block apartment buildings that housed the plant's workers and their families. Even in the larger cities, the principal employer was often a military production plant. In the city of Kharkov, the second-largest city in Ukraine, for example, a tank factory dominated the city's economy.

It was not enough, therefore, for the Russian Army to be withdrawn from the Warsaw Pact countries of Eastern Europe. The Russian troops returning home could not be easily re-integrated into a civilian economy because a civilian economy largely did not exist. An entirely new economy would have to be created out of the conversion of the defense industry and the demilitarization of the FSU.

Some key elements of the demilitarization of Russia included nuclear weapons dismantlement, the safeguarding of nuclear materials, and engaging scientists in the Russian nuclear weapons laboratories in productive collaborations that have potential commercial applications. In all these endeavors American scientists were integrally involved alongside Russian scientists.

The Russian nuclear weapons complex was in the 1990s larger and much more extensive than the American one. It consisted of two design laboratories (as does the American), along with 19 specialized production plants under the direction of the Ministry of Atomic Energy (MINATOM), seven of which were located in Moscow. The complex included everything from nuclear reactors to make the tritium and plutonium used in nuclear weapons, to plants that engineer and manufacture specialized components. In addition, scientific institutes and facilities in other republics that were formerly part of the Soviet Union (in, for example, Kiev, Kharkov and Minsk) did work related to the nuclear weapons program. The Soviet Union also maintained two sites for underground nuclear testing (compared to the one American test site in Nevada). At the peak of underground nuclear testing in the 1970s, the Soviet Union was testing at a rate approximately twice that of the three Western nuclear weapons powers (U.S., U.K., and France) combined.[2] Before the START I arms control agreement became effective in 1994 the Soviet

Union maintained some 10000 to 20000 nuclear weapons. This vast enterprise employed tens of thousands of people, including many of the country's best scientists. The conversion of this vast enterprise was not a trivial undertaking. The success of this conversion effort has depended on the final and permanent ending of the Cold War.

The first key element of the demilitarization, and a key component of compliance with START I, is the dismantlement of nuclear weapons. These devices typically contain between 5000 and 10000 individual parts and components, including fissile material like plutonium that is toxic and radioactive. Being radioactive means that the nucleus of the plutonium atom is unstable and hence spontaneously decays into a different nucleus with the emission of smaller energetic particles and high-energy photons. These emitted particles and photons are energetic enough that they can ionize other atoms that they strike. Thus, these ionizing radiations emitted from radioactive materials like plutonium can damage living tissue when a human being or other living organism is close enough to be exposed to them. Dismantling nuclear weapons is thus not only a very complex undertaking, it is also a dangerous and hazardous one. American and Russian scientists have worked jointly to develop safe dismantlement technologies.

Then there is also the problem of stabilizing, immobilizing, and disposing of the plutonium. The U.S. government-funded a program designed to help the Russians develop and implement industrial-scale immobilization technologies for plutonium from nuclear weapons dismantlement. Among the technologies that underwent development by American scientists is vitrification, in which the radioactive material is encased in a glass, or in a ceramic material that contains substances that absorb the neutrons that are emitted by the plutonium. The plutonium-bearing glass or ceramic is then sealed in a stainless steel can, and the sealed can is then placed in a larger steel canister that is filled with waste glass, making it impossible for anyone to reclaim the plutonium for use. The steel canisters are then buried in a safe geologic repository. Other technologies have also been studied and characterized in terms of their efficiency, cost, and non-proliferation benefits, that is, their potential to prevent the spread and re-use of the plutonium for nuclear weapons.

The Russians also pursued technologies for converting the fissile materials from nuclear weapons dismantlement to reactor fuel. Instead of safely immobilizing the radioactive material and disposing of it in a deep geologic repository, the material is "burned up" in a nuclear power plant. This was initially the preferred Russian option for fissile materials disposition, but it does not eliminate the need for safe disposition of nuclear waste products.

In addition, there is the more general problem of safeguarding all the special nuclear materials in Russia to prevent their falling into the hands of foreign nations or terrorist groups who could use them for the manufacture of rudimentary nuclear weapons. American scientists continue to work with Russian scientists to secure weapons-usable nuclear materials at all the sites that store, process, or use these materials. This includes not only sites within the nuclear weapons complex, but any site at which such nuclear materials are present. Nuclear-powered naval vessels, for example, also require protection technologies. These technologies include everything from locked covers for the storage racks for nuclear reactor fuel rods to TV monitors to special sensors placed at compartment entrances. American scientists continue to develop new advanced sensors for detecting the characteristic radiations from radioactive materials. Sensors have been installed at Russian airports, seaports, and various other border crossings in order to detect and intercept any clandestine shipments of nuclear materials out of the country.

Modern technologies for the protection and control of weapons-grade nuclear materials began being installed at 53 sites in the FSU, including three plutonium processing plants, two uranium enrichment sites, and the two Russian nuclear weapon design laboratories. New physical site protection and control technologies include video camera surveillance, portal radiation monitors, vault monitoring and surveillance technologies, and modern material inventory systems and tracking procedures. The idea is to move away from the old Soviet security model of "guns and guards" to a new security model based on modern security technologies that have already been developed and implemented in the United States.

The interdiction of smuggled nuclear materials presents a particular technical challenge. The technical problem is how to detect weak sources of the characteristic radiation from fissile materials in the presence of a high level of natural "background" radiation. A number of new technologies have been developed or are under development, including advanced sensors to image the very-high-energy photons, or gamma rays, that are emitted in the radioactive decay of uranium and plutonium.

Protecting just the nuclear materials, however, is not enough. The principal asset requiring protection in Russia is the large number of scientists formerly employed in the nuclear weapons enterprise who have special skills and knowledge. It would be a real threat to the security of both the U.S. and Russia if these skills and knowledge were to be made available to unfriendly nations or terrorist groups. The real success of the defense conversion effort depends on redirecting this workforce into productive non-weapons work. Accordingly, the U.S. government, as well as private financiers, have made money available to fund collaborative research projects between U.S. scientists and Russian scientists, and to develop commercial ventures for Russian scientific institutes. What was surely inconceivable prior to 1990 is now a reality: scientists from the Russian nuclear weapons labs visit the U.S. nuclear weapons labs, and vice versa, and a large and variable fraction of scientists at the Russian nuclear weapons labs are employed in collaborative non-weapons programs with American scientists. These collaborative programs provide opportunities to advance scientific understanding while contributing to the security and non-proliferation goals of both nations.

The U. S. government funds some, but not all, of these collaboration and commercialization efforts. Some money has been made available through a funding mechanism drafted by former U. S. Senators Sam Nunn (a Democrat from Georgia) and Richard Lugar (a Republican from Indiana).[3] The Cooperative Threat Reduction Program that was set up by the Nunn-Lugar Act of 1991 is still going as of this writing, but funding has been cut every year since 2009; starting in President Obama's second term in January 2013, prospects for continued cooperation between the U.S. and Russia on nuclear threat reduction have looked increasingly bleak. This will be discussed in more detail in the next chapter.

In addition, an intergovernmental organization called the International Science and Technology Center (ISTC) was established in 1992 by agreement of the Russian Federation, the U.S., the European Union, and Japan. The ISTC, with headquarters in Moscow, reviews proposals of scientific collaborations, and decides which proposals receive the funding that is provided by the member governments to the ISTC. The ISTC thus provides Russian scientists with opportunities for re-directing their scientific talents away from Cold War weapons work. U.S. participation in the ISTC is coordinated by the U.S. State Department. American scientists serve in a technical advisory capacity to the U.S. State Department to provide peer review of proposals submitted to the ISTC. Successful proposals are typically proposals that can show some clear prospects for commercialization of the proposed work, or are backed with partial funding from a private commercial source.

Other funding has been made available through a special

program set up by the American financier George Soros. This Soros funding mechanism works in much the same way as does the ISTC, accepting and peer-reviewing detailed technical proposals from Russian and other FSU scientists, in much the same way American scientists submit proposals to, say, the National Science Foundation or the National Aeronautics and Space Administration in the United States.

In total, it has not been a very large sum of money, compared to other defense spending, that has been made available to Russian scientists, but the money has been well leveraged to good effect, providing useful and interesting scientific work for a large number of former weapons scientists in the FSU, as well as for other scientists at Academy of Sciences institutes, including the Lebedev Institute in Moscow where Sakharov once worked.

Further scientific developments in the 1980s and 1990s converged with the political developments to accelerate the Cold War stand-down. The convergence of both these scientific developments and the political developments made possible the cessation of nuclear weapons testing, a particularly significant event in ending the Cold War.

The great bulk of the nuclear weapon tests that were conducted by both the United States and the Soviet Union were tests of new weapon designs. Tests were required because the physical processes that take place during a nuclear explosion, in which very extreme states of matter are generated, cannot yet be calculated from first principles with sufficient accuracy. This is because scientists do not yet know enough about the behavior and characteristics of matter at very high pressures, temperatures, and densities. Theories could not be checked against laboratory experiments because there had been no way until more recently to create the extreme material conditions that are present in a nuclear explosion in a controlled laboratory setting. The only way to certify that a new design would work as specified was to test it. This is largely what scientists did to certify weapon designs. Data from these tests were used to provide various "correction factors" in computer code simulations of nuclear explosions, but often the "correction factors" were different for different systems, providing a less-than-complete overall understanding of the basic underlying physical mechanisms. That is to say, nuclear weapons tests were generally tests of the integrated behavior of a very complex system, not controlled laboratory experiments meant to isolate individual physical processes.

When President George H. W. Bush decided in the early 1990s that no new designs would be produced, there was then much less need for nuclear testing. The specific nuclear weapons that remain in the stockpile have already been certified to work. Without nuclear testing, however, there remains the problem of how to predict the behavior of stockpile systems as they change with age. This is, in many ways, a much more challenging scientific problem than the problem scientists faced during the days of nuclear testing of certifying that a new design will work as it was designed to work. Nuclear weapons were meant to have only a limited lifetime in the stockpile, usually about ten years or so. After that, they would be phased out and replaced with more modern designs. If there are no new designs being produced, and the country continues to rely on an aging nuclear weapons stockpile as its ultimate military deterrent, then some way must be found to guarantee the safety, reliability, and military effectiveness of the stockpile weapons over many decades of stockpile life.

The problem is not unlike that of, say, closing all aircraft production plants, grounding the entire fleet of airliners, stopping all aircraft design and full-scale wind-tunnel tests, while at the same time trying to assure that if an airplane is needed thirty years from now that one of the aging craft can be taken out of the hangar and that it will take off and fly.

Fortunately, what has made this problem somewhat tractable has been the development of experimental capabilities to create matter at extreme conditions in laboratory devices, contain it, and measure its properties. These developments in experimental capabilities have taken place in parallel with developments in basic understanding of matter at extreme conditions, along with advances in computational capabilities as computer technology has improved.

All matter consists of a collection of atoms. All atoms, as we have discussed, consist of a positively charged nucleus, containing protons and neutrons, surrounded by negatively charged electrons in discrete orbits around the nucleus. By treating each electron in the atom as a wave, with a wavelength that depends on its momentum (the product of its mass and its velocity), one can use the equations of quantum mechanics to figure out, in principle, the exact structure of each atom, that is, how the electrons are arranged within the atom. The electronic structure of an atom determines its physical and chemical properties. Quantum mechanics is successful in explaining, for example, how a sodium atom is different from a carbon atom, and why the chemical bonds these atoms form with other atoms are different.

One can also use the laws of quantum mechanics to figure out how atoms are held together by electrical forces and arranged in a solid. If we take a solid material, a lump of water ice, say, and heat it up, the added thermal energy will make the atoms in the solid lattice vibrate, or jiggle, more vigorously. Eventually, a temperature will be reached at which the thermal energy of the vibrations will be greater than the electrical energy holding the atoms together in the solid lattice, and the material will melt. We say that it undergoes a change of phase from a solid state to a liquid state. The lump of ice becomes a puddle of water. Along with the change of state comes a change in many of the material's characteristic properties. The liquid loses its resistance to changes in shape, will take on the shape of whatever container it is in, and can flow in response to external forces on it. Add yet more energy to the liquid and it undergoes yet another change of phase to a gaseous state. The water becomes steam. As a gas, the atoms are no longer bound together, and interact only upon collision with one another.

If we add still more energy to the gas eventually the thermal energy of the atoms will exceed the energy that binds the outermost electrons to the atom's nucleus. The atoms then become ionized, that is, one or more of the electrons become detached from the atomic nucleus, and the material undergoes yet another change of phase from the gaseous state to the plasma state. A plasma is simply a fluid of electrically charged particles. The overall plasma may be electrically neutral, containing an equal charge of negatively charged electrons and positively charged nuclei, but unlike a gas, in which the particles only interact when they collide with each other, each particle in a plasma can interact with all the other particles in the plasma by the electric force that exists between charged particles. The plasma can thus display a whole range of new collective behaviors that are not observed in the other states of the matter.

Plasma behavior can be extraordinarily complex. Plasmas can flow like fluids and can also be strongly bound together like solids. Since moving electric charges emit and absorb electromagnetic radiation, the interaction between plasma particles and radiation plays a very large role in the material's overall behavior.

Most matter in the universe is in a plasma state. The Sun and all stars are plasmas. Much of the interstellar gas in galaxies is also in the plasma state. Plasmas with which we are familiar in everyday life include the glowing gas in a fluorescent light bulb. Exploding nuclear weapons are also plasmas.

Indeed, the matter in an exploding nuclear weapon starts out as a solid and goes through very many phase changes over a very large range of temperature, pressure, and density conditions. Plutonium, in particular, is a very complicated substance, because it undergoes several solid-to-solid phase changes when it is heated by only a small amount. This happens because the atoms of plutonium rearrange themselves in different ways at different temperatures, and the properties of these different solid states can be quite different from one another. For example, the volume of a particular amount of plutonium can increase by as much as one-fifth when it is heated by 400 degrees. Since a deployed nuclear weapon is exposed to environmental conditions that span a range of temperatures nearly this large, such changes in basic dimensions are a big concern.

Adding to the complication, plutonium is a radioactive material. It decays by emitting helium nuclei. Helium, which is a gas, collects within the material and can cause defects and damage spots that can affect the explosion behavior. In addition, hydrogen gas can evolve from various hydrogen-bearing compounds within some of the other materials, including from water vapor in the air. Hydrogen, when it chemically reacts with metals, can lead to corrosion. A large number of other aging effects can also be present, and some or all of these aging effects can result in defects that can affect the design performance of the device.

In order to maintain confidence in the reliability of aging nuclear weapons, it is important to be able to predict the effects of these age-induced changes on the explosion. This requires having the ability to do controlled laboratory experiments on matter at extreme conditions to test our understanding of the basic physical mechanisms and how they are affected by changes in initial conditions.

Fortunately, the development of a large variety of devices to create and control plasmas, and diagnostic instruments to measure their properties and behaviors, has made such experiments possible. Indeed, miniature nuclear fusion explosions can be created in a controlled laboratory setting by using very high-power lasers. The laser beams are focused onto a target which consists of a capsule containing the fusion fuel, usually gaseous deuterium and tritium, which are isotopes of hydrogen. The energy absorbed by the capsule shell from the laser beams implodes the shell, causing the fuel inside it to be heated by compression to the temperature, pressure, and density conditions required for the fusion reactions to take place. The laser does not create a mini nuclear weapon. There is no laser with enough energy to implode a critical mass of fissile material. The fissile material in a nuclear weapon is imploded by conventional high explosives. The implosion assembles the fissile material into a "critical" state where the fission reactions "runaway," producing more energy than the energy expended by the conventional explosives in assembling the fissile fuel. To reach this critical state requires a certain minimum mass of fissile material. This critical mass is much larger than what can be imploded by even the most energetic lasers that can be built. Thus, in laser fusion, there is no fission and no fissile material. The laser is used instead to create the conditions in which fusion, not fission, reactions can take place. It is not possible with current lasers to get more fusion energy out than the energy the laser puts in. Thus, laser fusion experiments do not provide a way to test miniature nuclear weapons. The real value of laser fusion experiments, however, to studying nuclear weapons is that they create the material conditions of temperature, density, and pressure in plasmas undergoing nuclear fusion reactions that are similar to the conditions in a nuclear weapon, so it is possible, in principle, to study the relevant physical mechanisms that affect the performance of an aging nuclear weapons stockpile in a controlled laboratory setting.

Concomitant developments in diagnostic instrumentation have made it possible to measure the conditions in fusion plasmas on the required very short time scales and the very small spatial scales. These diagnostic advances have been greatly aided by advances in digital electronics technology, as well as advances in chemistry and materials science that have allowed the development of new detector materials for the high-energy photons, like x-rays and gamma rays, and elementary particles like neutrons and protons, that are emitted by fusion plasmas. There have been many parallel advances in mathematical and computational science, too. New algorithms that can solve the complex coupled set of mathematical equations that describe the motion and behavior of fluid plasmas have been developed at laboratories and universities all over the world. These developments, coupled with the extremely rapid advances in computer speed and memory capability over the years since the first commercially available computers, have opened a whole new field of scientific investigation using "numerical experiments." In "numerical experiments" a scientist can conduct virtual experiments on his or her computer, figuring out how a complex plasma system, for example, behaves under the influence of certain external conditions, and how the behavior changes in response to changes in the external conditions. Such numerical experiments, validated by real laboratory experiments, have led to numerous breakthroughs in recent years in understanding complex behavior of matter at extreme conditions.

All of these developments, collectively, have made possible what was once unthinkable: the maintenance of a nuclear weapon deterrent capability in the absence of nuclear testing. This has given new life to the old goal of concluding a treaty banning nuclear weapons testing.

The idea of stopping nuclear weapons testing as the first step in eliminating nuclear weapons – or stopping the development of new ones – is not a new idea. Nor is it a partisan one, since it is an idea that has been embraced by both Republican and

Democratic Presidents. As early as 1960 President Dwight Eisenhower expressed the belief that a ban on all nuclear weapon testing would be a good thing. In his Annual Message to the Congress on the State of the Union that year he called for pursuit "of negotiations looking to a controlled ban on the testing of nuclear weapons." His successor, President John F. Kennedy, shared President Eisenhower's passion in wanting an end to nuclear weapons testing. He was not able to achieve his stated goal of "assuring the end of nuclear tests of all kinds,"[4] but he did succeed, as discussed in Chapter 4, in concluding a bilateral Limited Nuclear Test Ban Treaty with the Soviet Union in 1963, which ended all nuclear weapons tests by the United States and the Soviet Union conducted in the atmosphere and in outer space.

It was President Bill Clinton thirty years later who with particular vigor pursued the goal of an international treaty ending all nuclear weapon testing and saw its actual accomplishment. Only about five weeks after taking office in January 1993 President Clinton directed a review of U.S. policy on nuclear testing and a Comprehensive Test Ban Treaty (CTBT). He agreed at a summit meeting with Russian President Boris Yeltsin on April 4, 1993, to begin negotiations on a multilateral CTBT. In July the President announced the completion of the Presidential Review he had ordered the previous March, extended the testing moratorium that had been started by President George H. W. Bush in 1992, and stated his intention to negotiate a CTBT. The following December the United Nations General Assembly for the first time ever passed a resolution in support of a CTBT. Negotiations began the following month at the Conference on Disarmament in Geneva, Switzerland.

Progress towards a CTBT was not without setbacks. Although both the United States and Russia have observed the voluntary testing moratorium in effect since late 1992, not all nuclear-weapon-capable states have done so. The People's Republic of

China resumed nuclear weapons testing in October 1993, just as negotiations on a CTBT were set to get underway. Then, in June 1995, just after the Conference on Disarmament had set itself a Fall 1996 deadline to complete the Treaty, France announced the resumption of nuclear weapons testing. French President Jacques Chirac told the world that France would conduct only eight tests, and then be ready to sign onto the CTBT by Fall 1996, but the United States nonetheless saw the French announcement as a setback to the negotiations, and issued a statement of regret about the French decision.

True to his word, though, President Chirac announced the end of French nuclear testing on January 29, 1996, after the eighth new test and less than five months after the first new test. Then, the People's Republic of China ended its nuclear testing and rejoined the test moratorium at the end of July. Draft treaty language was prepared, and after some compromises were reached on some difficult issues, it once again looked like the Fall 1996 deadline would be met. Less than a month before the United Nations General Assembly meeting was to convene on September 10 and adopt the CTBT, however, the negotiators in Geneva issued a report stating that "no consensus" could be reached on the text of the CTBT or even on submitting it to the full Conference on Disarmament, largely because of objections by India. This seeming setback did not stop the Australian Foreign Minister, Alexander Downer, from announcing that his country would nonetheless submit the draft CTBT to the United Nations General Assembly for endorsement and signature.

The Australian resolution was indeed adopted by the United Nations General Assembly on September 10, 1996, by a vote of 158 to 3 (India, Bhutan, and Libya were opposed), with five abstentions (Cuba, Lebanon, Syria, Mauritius, and Tanzania). President Clinton became the first head of state to sign the new Treaty, for the United States, on September 24. One year later when President Clinton again addressed the United Nations

General Assembly, he referred to the CTBT as "our commitment to end all nuclear tests for all time – the longest-sought, hardest-fought prize in the history of arms control. It will help to prevent the nuclear powers from developing more advanced and more dangerous weapons. It will limit the possibilities for other states to acquire such devices. – Our common goal should be to enter the CTBT into force as soon as possible, and I ask for all of you to support that goal."

As of February 2019, 185 nations had signed the CTBT, including all five declared nuclear weapon states at the time the treaty was first opened for signature (U.S., U.K., France, Russia, China). One hundred sixty-eight of these countries have also ratified the Treaty. The Treaty specifies which 44 countries must not only sign, but ratify the Treaty in order for it to go into effect. Of these 44, 36 have so far done so. Among the remaining eight, three have not even signed (India, Pakistan, and North Korea); the other five, who have signed the Treaty but not ratified it, are China, Egypt, Iran, Israel, and the United States. The failure of the United States to ratify the CTBT has become a major obstacle to the final resolution of the Cold War. More will be said about this in the next chapter.

Nonetheless, preparations have begun for administering and verifying compliance with the treaty once it goes into force. The technical provisions for treaty monitoring and verification were among the decisive factors in the adoption of the CTBT by most of the world's countries. The verification technologies, developed by scientists in the United States, in Russia, and in other countries, gave credibility to the claims of Treaty proponents that the Treaty could be adequately monitored for compliance. American and Russian scientists, in particular, had, by the time of the adoption of the CTBT, a long history of developing treaty monitoring and verification technologies. Indeed, a sophisticated verification regime was included as part of START I, the first arms control treaty, as discussed in the last chapter, to call for actual reductions in nuclear warheads and missile systems. START I was in fact designed right from the start with verification in mind.

START I calls for both intrusive and non-intrusive monitoring, and also contains provisions prohibiting interference with monitoring technologies. The main component of the non-intrusive monitoring, what are called "national technical means," are surveillance satellites. These "spy" satellites contain a variety of advanced sensors to allow each side to observe what the other side is doing on its own territory with respect to, say, moving and testing missiles. In addition, each country uses its own instrumentation to record the flight test data sent back to ground stations during missile test flights. START I prohibits any encryption or concealment of missile flight test data, including encapsulation and jamming, and obliges each side to make available to the other side relevant telemetry and interpretative data for each test flight, including the missile acceleration histories.

More intrusive monitoring of START I consists of installed instrumentation to provide continuous monitoring at the perimeters and portals of each side's assembly facilities for mobile intercontinental ballistic missiles. The Treaty also allows twelve types of on-site inspections, and formal mechanisms to raise compliance concerns.

This experience with START I monitoring and verification was valuable in designing verification provisions for the CTBT. These provisions are encompassed in the International Monitoring System (IMS). The IMS is a global network of monitoring stations set up to detect clandestine nuclear weapons test explosions. There are four principal monitoring technologies that scientists have developed to detect such explosions. Three of the four detect the characteristic wave motions that the explosion induces in its surroundings. One such wave motion is the shock wave produced in the solid earth for underground explosions. An underground explosion produces two types of wave motions, a compression wave in which the soil and rock are moved back and forth in the direction the wave travels, and a shear wave in which the soil and rock are moved back and forth across the direction the wave travels. An underground explosion produces much more compression wave motion than shear wave motion, and thus can be distinguished from an earthquake, which typically produces much more shear wave motion. Other sensors are designed to detect the characteristic sound wave produced in an underwater explosion, and still others can detect the low-frequency sound waves that travel large distances in the atmosphere from an underground explosion. The fourth type of detector technology allows detection of the characteristic radioactive debris that is produced from an atmospheric explosion, or an underground explosion from which the hot, radioactive gases escape.

There are a sufficient number of monitoring stations containing sensors of sufficient accuracy to be able to detect clandestine nuclear explosions with reasonable certainty almost anywhere on earth. It was the combination of scientists' development of these technologies and the advances discussed earlier in being able to compute and experiment with the behavior of matter at extreme conditions that made the CTBT possible.

Further, even while dismantlement according to the provisions of START I was underway, further reductions in nuclear weapon arsenals of both the United States and Russia were negotiated in START II. The negotiations leading to this follow-on agreement got underway following the summit meeting between Presidents George H. W. Bush and Boris Yeltsin in June 1992. The two presidents met again in Moscow and signed START II on January 3, 1993. START II called for the elimination of all multiple independently targeted missile systems (i.e., missiles carrying several nuclear warheads, all of which can be directed at separate targets), and an overall further deep cut in the total numbers of nuclear weapons on each side. The plan was that by 2003 or so, both countries would be left with 3000 - 3500 nuclear weapons each, down from a few tens of thousands each when START I went into force and the Soviet Union dissolved in 1991. There were several new developments that took place in the early 2000s, however, that led to the end of START II. These developments are discussed in the next chapter.

Meanwhile, the United States and its Western allies, U.K. and France, embarked on a program to maintain the safety and reliability of their remaining aging stockpile weapons. The U.S. "stockpile stewardship" program[5] is aimed at maintaining the U.S. nuclear deterrent in the absence of nuclear weapons testing. It has several key components. One main component is a basic scientific research component, the purpose of which is to expand our basic knowledge about the behavior of matter at extreme conditions. Instead of money being directed to the design, development, and manufacture of new nuclear weapons, money is directed at several major new scientific initiatives to provide the infrastructure for the scientific research at the core of stockpile stewardship. The largest, and costliest, of these scientific initiatives is the National Ignition Facility (NIF), a large high-power laser facility that has been constructed at the Lawrence Livermore National Laboratory located in California.

The NIF Project was a controversial one right from the start, along with the entire Stockpile Stewardship Program of the U.S., and remains so to this day. NIF was plagued during its early phases by some technical difficulties, management problems, and serious cost overruns. Critics maintained that NIF was too costly, unlikely to achieve its stated scientific objective of achieving thermonuclear ignition in a controlled laboratory setting,[6] and that it will be used to design and test new nuclear weapons, in violation of the CTBT.[7] Proponents, on the other hand, countered the technical arguments of the critics with scientific studies showing that the scientific basis for ignition with NIF was sound. Numerous technical reviews of the project have reached the same conclusion.

The real argument, though, is a political one. The entire Stockpile Stewardship Program is a grand compromise between the right and the left. The right, still committed to the idea of military dominance, wants not only a nuclear weapons program, but a return to nuclear weapons design and testing. For them, the Cold War is not over, just temporarily on hold, or even re-ignited, as discussed in the next chapter. They are skeptical of Russia's commitment to a democratic transition (with some justification), and think that the U.S. should be prepared for a resumption of Cold War hostilities. Many members of Congress on the right, however, have been willing to live with no nuclear testing as long as a viable nuclear weapons capability can be maintained via a Stockpile Stewardship Program. They have thus been willing to fund the program even if it does not give them everything they want.

The left, meanwhile, wants no nuclear weapons program at all. Still committed to the idea of unilateral disarmament, they see no need for any nuclear deterrent when there is no longer a threatening Soviet Union to deter. Many members of Congress on the left, however, have also been willing to fund a Stockpile Stewardship Program if this is the price for obtaining the CTBT, and with it, a permanent end to nuclear weapons testing. Thus, both sides get something they want, but it is a fragile compromise. Like all compromises, it must be maintained and nourished over a long period of time, even as the political leaders responsible for the decisions change with time. In practice, this is a very difficult thing to do. It was made even more difficult during the last years of the Presidency of Bill Clinton, because partisan and ideological differences between the two political parties in the United States erupted into open political warfare with the impeachment and trial of the President on charges that he perjured himself in testimony concerning an alleged sexual liaison with a female member of his staff. It was not too long after President Clinton survived the impeachment vote in the Senate, and with partisan bitterness on both sides still running high, that the President called on the Senate to ratify the CTBT, nearly three years after he first submitted it to the Senate for their consideration for ratification. Then, the fragile compromise began to unravel.

Just as ideological rigidity on both sides of the Cold War conflict had prolonged the conflict, ideological rigidity now threatens reaching a final and permanent ending of the Cold War. And just as it was the scientists on both sides who provided the pathway to ending the Cold War, it could be scientists again who guide us down the path to a final and permanent end of the Cold War.

1. Yeltsin was elected President of the Russian Soviet Federative Socialist Republic, one of the 15 republics that composed the Soviet Union, on June 12, 1991, taking office on July 10. When the coup was launched the following month against Gorbachev, Yeltsin quickly emerged as a leader of the anti-coup forces, and a leading proponent of Russian nationalism and independence. When the Soviet Union dissolved in December of that year, Yeltsin became the President of the newly independent Russian Federation. He was elected to a second term in 1996.
2. Samuel Glasstone, editor, *The Effects of Nuclear Weapons*, published by the U.S. Atomic Energy Commission, April 1962, and available through the U.S. Government Printing Office, Washington, D.C. 20402.
3. The Soviet Nuclear Threat Reduction Act of 1991, 22 U. S. Code Section 2551, colloquially referred to as the Nunn-Lugar Act after its two co-sponsors in the U. S. Senate. The Act was an amendment to the Arms Export Control Act. It

was passed in both the House and the Senate in November 1991 and signed into law by President George H. W. Bush on December 12, 1991. Title II of the Act provides that the U. S. will cooperate with Russia on the storage, transportation, dismantling, and destruction of Soviet nuclear weapons. Implementation was via the Cooperative Threat Reduction Program, run by the Defense Threat Reduction Agency, a semi-autonomous agency within the U.S. Department of Defense.

4. President's address to the United Nations General Assembly, United Nations, New York, New York, September 25, 1961.
5. "Science-Based Stockpile Stewardship," by Raymond Jeanloz, *Physics Today* Vol. 53/12, pages 44-50, December 2000.
6. See S. E. Bodner, *Journal of Fusion Energy Vol. 11(2)*, page 73, 1992; see also *Science* Vol. 277, page 304, 1997.
7. See *Science* Vol. 289, pages 1126-1129, 2000.

Chapter 10

The Aftermath, Part II: Unraveling

One factor that has frustrated the Cold War stand down is Russia's so-called "nationalities problem." This problem stems from the fact that the Soviet Union was the inheritor of most of the territory that had been part of the Russian Empire prior to the First World War. Stalin, after he came to power in 1923, had organized this vast territory into 15 separate Soviet Socialist Republics, partially (but not completely) along ethnic or linguistic boundaries. This organization was partly a recognition of the fact that the Soviet Union was not a single ethnically homogeneous nation, but a federation of many different and distinct "nationalities," held together only by the totalitarian control of the Soviet Communist Party. The organs of state control were centered in Moscow, which doubled as the capital of the Soviet Union and the capital of the Russian Soviet Socialist Republic.

When the Soviet Union dissolved in December 1991, and the totalitarian control of the Soviet Communist Party no longer existed, these 15 Soviet Socialist Republics became independent countries. It was only one of these newly independent countries – the Russian Federation – with its capital still in Moscow, that ended up inheriting all the organs of government and state control that had previously been those of the Soviet Union. These included, of course, the armed forces and the Soviet Union's stockpile of nuclear weapons, wherever these forces and weapons were located.[1] How the new Russian state related to and interacted with the other 14 newly independent republics, especially since all of them had some ethnic Russian population, some of whom felt alienated from their mother country, became complicated and contentious. These complications manifested in different ways in different places, and Russia had different responses to the different manifestations.

To summarize these complex developments and how they contributed to the unraveling of the Cold War stand down, let us consider the 14 newly independent republics in three regional groupings.

In Central Asia, the five "Stans" (Turkmenistan, Uzbekistan, Kirgizstan, Tajikistan, and Kazakhstan) had the smallest population of ethnic Russians. All the Stans were, and still are, majority Muslim, but ethnically and linguistically very different from one another. The boundaries of the Stans, drawn by Stalin in the 1920s, did not coincide exactly with ethnic boundaries. For example, there is a sizeable Tajik population in Uzbekistan, the two ethnic groups speaking entirely different languages.[2] Under the centralized command of the economy in the Soviet Union, each Stan was separately assigned to organize its economy around one product. For Turkmenistan, for example, it was natural gas; for Uzbekistan it was cotton. Thus, after independence not one of the Stans had a sufficiently diversified economy that allowed it to transition easily to a viable market economy. The Stans also had political spill-over effects of the long Soviet military presence in neighboring Afghanistan. All of these problems led to different outcomes for the five different Stans, with Turkmenistan, for example, becoming the richest of the Stans, because of its vast natural gas resources, but a dictatorship with strict police-state controls. Uzbekistan, on the other hand, initially under authoritarian rule, eventually succeeded in making a transition to a more democratic and open society. Tajikistan was beset initially with a civil war and an Islamic insurgency. Kirgizstan is still struggling with ethnic conflict.[3] Miraculously, no wars broke out between the Stans, or between any of the Stans and Russia. Russia has also avoided direct military interference in the internal ethnic conflicts in the individual Stans.

This was not the case, however, in all 6 of the republics on the southwest periphery of Russia. The first of the former Soviet republics to engage in armed conflict with each other were Armenia and Azerbaijan. The original Soviet-defined border between these two republics left a large Armenian minority population in Azerbaijan, mostly in the Nagorno-Karabakh region. Shortly after independence in 1991, Armenia invaded Azerbaijan to recover this region. By the time a Russian-brokered cease-fire was agreed to in 1994, 30,000 people had died, one million were displaced, and the Armenian economy was in shambles because of the blockade of this landlocked country by both Azerbaijan on their eastern border and Turkey on their western border. Even though Armenia captured about 16% of Azerbaijan's territory in the war, including the disputed region, there is still no internationally recognized border between the two countries and no peace agreement. Armenia has since made the transition to a market economy, but the country still suffers economically in the absence of peace with its neighbors to the east and west. They have, however, been able to maintain trade with the outside world because they maintain good relations with their neighbor to the north, Georgia, and their neighbor to the south, Iran.

It was in the former Soviet Republic of Georgia, however, that Russian military intervention in an internal conflict began to re-ignite the Cold War with the U.S. The history of Georgia since independence in 1991 has been a complicated one of on-again, off-again instability and turbulence, including a violent coup, and ethnic warfare caused by tensions between Russian separatists in Abkhazia and South Ossetia – regions in Georgia that were considered autonomous in the Soviet Union – and the majority Georgian population. The simmering tensions led to open warfare in 1992-1993 that displaced some 270,000 Georgians by separatists in the two regions. Russia supported the separatists. Even though an agreement was reached between

Georgia and Russia in 2005 that withdrew Russian troops from two of the three military bases they maintained in Georgia, continuing tensions between the two countries led again to war in 2008. French President Nicolas Sarkozy brokered a cease-fire agreement in August 2008. Two weeks after the cease-fire, however, Russia recognized both Abkhazia and South Ossetia as independent republics, and Georgia responded by severing diplomatic ties with Russia, considering the two regions as Russian-occupied Georgian territories.

The European Union and the U.S. were alarmed by the Russian invasion of Georgia, with President George W. Bush threatening political and economic isolation of Russia if they did not withdraw their troops from Georgia under the terms of the ceasefire.[4] Actions taken by the U.S. included using U.S. military transports to return Georgian forces from Iraq to defend the Georgian capital, Tbilisi, from Russian attack; along with European allies, raising financial aid for Georgia; and levying sanctions against the separatist regions.[5] No stronger measures were taken, largely because U.S. foreign and defense policy was focused on the wars in Iraq and Afghanistan. Nonetheless, the continuing conflict between Russia and Georgia set the stage for a deeper worsening of the U.S.-Russia relationship a few years later with the Russian military intervention in Ukraine, another of the newly independent former Soviet republics. But we are getting a little ahead of the story.

In the other two newly independent former Soviet republics on Russia's southwest periphery, Moldova and Byelorussia, pro-Russian leadership guaranteed them a future free of armed Russian intervention.

The three Baltic republics formerly part of the Soviet Union – Estonia, Latvia, and Lithuania – presented yet another front in the re-igniting of Cold War tensions between the U.S. and Russia. Here the issue was the expansion of the North Atlantic Treaty Organization (NATO) into former republics of the Soviet

Union.

NATO is a mutual defense alliance between the U.S., Canada, and several European countries. It was established in 1949 by the North Atlantic Treaty, and at its founding included ten West European countries.[6] Article 10 of the Treaty lays out the provisions and procedures for adding new member states to the alliance. After the end of the Cold War and the dissolution of the Soviet Union in 1991, there was a lot of political pressure on both sides of the former East-West divide to expand NATO eastward to include not only the former Warsaw Pact countries of Eastern Europe, but also some of the republics of the former Soviet Union.[7] Russia opposed these moves at expansion, seeing them as a direct threat to their own security. Russia's worst fears were realized when, at the 2002 NATO Summit in Prague seven new eastern European countries were invited to join NATO, with formal membership being granted at the 2004 NATO Summit in Istanbul. Among the new member states joining NATO were the three Baltic republics that were formerly part of the Soviet Union. The three countries became the first – and so far, the only – former Soviet republics to join NATO.

The effect on Russia of Estonia, Latvia, and Lithuania becoming NATO member states was simply to harden its resolve to oppose, with force, if necessary, other FSU republics being swept into NATO. After 2003, Georgia began to seek closer ties with NATO in their ongoing conflict with Russia over their breakaway regions. Fear of Georgia joining NATO was one factor in the Russian military invasion of Georgia in 2008 discussed earlier. Russian fear was further intensified, and U.S.-Russia tensions increased, when Georgia was named a NATO "aspirant country" at the North Atlantic Council meeting on December 7, 2011.

The tensions over NATO expansion into former Soviet republics have fortunately not led to any armed conflict between NATO forces and Russia, at least not as of this writing (2020).

NATO forces, however, have intervened in the Balkan Wars against forces closely allied with Russia.

The Balkan Wars arose in the 1990s out of independence movements in all six of the member republics of Yugoslavia. The country of Yugoslavia was a creation of the country's founder and long-time leader, Josip Broz Tito, who forged the nation as a mini-Soviet Union, under the control of the Yugoslav Communist Party, from the ruins of the six separate republics left from the Second World War. The Yugoslav union was dominated by Serbia, with its capital in the Serbian capital of Belgrade. After Tito's death in 1980, ethnic tensions rose, and the separate republics began to rebel against rule by the dominant Communist regime in Serbia. One by one, they began to peel away from Yugoslavia, starting with Slovenia and Croatia in 1990 and 1991. The attempt by Bosnia-Herzegovina to split off from Serbia led to a brutal civil war that raged for three years (1992-1995), with Bosnian Serbs establishing a separate political entity within Bosnia-Herzegovina that remained in federation with Serbia. NATO intervention in the war targeted the Bosnian Serb army. This intervention alienated both Serbia and its close ally, Russia.

Then, NATO launched a massive bombing campaign on Serbia itself in 1999 to stop the brutal suppression of the independence movement in the Serbian province of Kosovo, a largely (greater than 90%) ethnic Albanian region in southern Serbia. These actions by NATO are likely the main reason why to this day both Bosnia-Herzegovina and Serbia are the only two republics of the former Yugoslavia that have not become NATO members.[8]

This brief historical review shows that the formal end of the Cold War did not really end the military stand-off between the U.S. and Russia; it just re-shuffled the alliances. NATO has expanded since 1991 to include not only some of the former eastern European states that were previously part of the

Soviet-led Warsaw Pact, but also some of the former Soviet republics. Meanwhile, some of the other former Soviet republics have allied with Russia in a new military alliance called the Collective Security Treaty Organization (CSTO). Created by a treaty that was signed on May 15, 1992, in Tashkent, capital city of Uzbekistan, this collective military alliance prohibits its members from joining other military alliances, such as NATO.

Relations between these separate military alliances – NATO and CSTO – have been further complicated by the fallout from the ethnic and political tensions stemming from the nationalities problem in both the former Soviet Union and the former Yugoslavia. In several of these places, these tensions have led to armed conflict. These armed conflicts have certainly hindered bringing the Cold War to a final and complete end.

Another consequence of these tensions and conflicts has been the beginning of the unraveling of the arms control agreements that had been so carefully crafted during the whole course of the Cold War. In the first decade of the new century, three important arms control agreements were undone, with long-term consequences yet to be known.

The first of the arms control treaties to die was the Anti-Ballistic Missile (ABM) Treaty. The ABM Treaty was, in its essence, the linchpin of the whole concept of nuclear deterrence. The strategy of deterrence was based on the idea that each side would be deterred from launching a strike with nuclear weapons on the other side because the other side would be able to retaliate and wreak massive destruction on the side that launched the first strike with the weapons that survived the first strike. This posture of "mutually assured destruction" could work only if neither side had defenses against attack by intercontinental ballistic missiles. In other words, if one side had such defensive capability, then they could, in principle, be emboldened to launch a first strike against the other side, having assurance that their missile defense would protect them from any retaliatory strike. Thus, the main motivation for the ABM Treaty was to discourage any side from launching a first strike.

The ABM Treaty was signed in Moscow on May 26, 1972, by U.S. President Richard Nixon and Soviet Premier Leonid Brezhnev. The initial terms of the treaty were to limit each side to having only two ABM sites. In 1974 a Protocol was added to the treaty language to further limit each side to only one missile defense site. Russia has their one site protecting Moscow. The U.S. has its one site protecting the ICBM complex in North Dakota.

The ABM Treaty worked for three decades to prevent nuclear war between the U.S. and Russia, and helped to make possible the phased reduction of offensive nuclear weapons in the other arms control agreements that were negotiated over the years, such as START I and START II discussed in the two previous chapters. There was yet another development in the early 2000s, however, that upset the whole arms control construct. It started on September 11, 2001.

After the Al Qaida terrorist group's attack on the U.S. on that date, followed shortly thereafter by the U.S. military invasion of Afghanistan, the whole focus of U.S. defense and foreign policy changed dramatically. In parallel with the new focus on countering terrorism, the U.S. and its allies were focused on the rising threat of a potentially nuclear-armed Iran. The Islamic Republic of Iran, which funded and provided material support to terrorist groups operating in the Middle East, was developing not only a nuclear weapons capability, but also had a missile development program underway. These missiles were thought to have a range that would allow them to reach Europe with nuclear weapons. In response to this perceived threat, the George W. Bush administration began the deployment of missile defense sites in the new NATO countries of eastern Europe. Russia saw this move as a direct violation of the ABM Treaty.

On December 13, 2001, President George W. Bush withdrew the U.S. from the ABM Treaty. The very next day the Russians responded by withdrawing from the START II Treaty.

START II, as discussed in the previous chapter, was a follow-on treaty to START I. It banned the deployment of multiple independently targeted reentry vehicles (MIRVs) on intercontinental ballistic missiles (ICBMs). MIRV capability allows one missile to carry several nuclear weapons, each of which can be directed to a different target. Thus, MIRV capability can be destabilizing to deterrence in much the same way as missile defense is, in making a first strike more likely.

This Treaty had been negotiated in 1991 and 1992, and signed by U.S. President George H. W. Bush and Russian President Boris Yeltsin in Moscow on January 3, 1993. It took several years, however, before the Treaty went into effect, because of long delays in the ratification process. It was not until January 26, 1996, that the U.S. Senate ratified START II by a vote of 87 to 4. The ratification delay in Russia was even longer, because of Russian protests in the Duma, Russia's legislature, over American military actions in Kosovo and NATO expansion. It was not until April 14, 2000, that the Russian Duma voted to ratify the treaty and it went into force.

Thus, START II was in effect for little more than two years before the Russian withdrawal. To this day, Russia still maintains a MIRV capability on their ICBM force. In contrast, the U.S. began a unilateral dismantlement of MIRV capability in October 2002, just four months after the Russian withdrawal from the treaty. This included withdrawal of the entire fleet of Peacekeeper ICBMs. This process was completed by September 19, 2005, leaving the Minuteman III as the only operational American ICBM, each missile capable, in principle, of carrying three MIRVs.[9]

The ABM Treaty and START II were not the only arms control agreements to fall by the wayside in the early 2000s. START III met the same fate.

START III was meant to be a continuation of the phased reduction of the strategic nuclear weapons arsenals, and their delivery systems, on both sides. The framework for negotiations on START III as a follow-on treaty to START I and START II was first discussed at a summit meeting between U.S. President Bill Clinton and Russian President Boris Yeltsin in Helsinki in 1997. Actual negotiations, however, were delayed for the same reasons that delayed the Russian Duma's ratification vote on START II, namely, Russian protests over the NATO bombing campaign in Serbia and NATO expansion, as well as Russian opposition to U.S. plans to deploy a limited missile defense system in Europe. Once negotiations did get underway, negligible progress was made, largely for the same reasons. Ultimately, START III negotiations were abandoned altogether, and attention shifted to a new approach to nuclear arms control.

The new approach focused on limiting the total number of operationally deployed nuclear weapons. This was a very different approach than what was done in the START negotiations. START sought to limit the number of weapons systems according to their means of delivery. Both nations deployed nuclear weapons on three different delivery vehicles: intercontinental ballistic missiles (ICBMs), long-range bombers; and submarines. Russia always had a very different mix of these three things than did the U.S. The largest fraction of Russia's nuclear weapons was – and still is – deployed on land-based ICBMs. For the U.S., however, submarine-launched ballistic missiles (SLBMs) were – and still are – a much greater fraction of the total. This difference between the composition of the two countries' arsenals greatly complicated START negotiations. This was because it was difficult to agree on cuts to the three legs of the triad that would not be perceived as a disadvantage for one side or the other. For example, land-based ICBMs are much more vulnerable to a first strike than are SLBMs. It was technical considerations like this that frequently complicated START negotiations. Thus, when the focus shifted to limiting just the total number of strategic nuclear weapons, irrespective of declared attribution to the delivery vehicle, both sides were able to come to an agreement on a Strategic Offensive Reduction Treaty (SORT).

SORT was signed in Moscow by U.S. President George W. Bush and Russian President Vladimir Putin on May 24, 2002. Once again, ratification was delayed in the Russian Duma, with a new reason for the delay being added to the reasons for Russia's withdrawal from START II: the U.S. military invasion of Iraq. Nonetheless, SORT finally became effective on June 1, 2003, with an expiration date set for February 5, 2011, a little more than one year after the expiration date for START I.

SORT limited the strategic nuclear arsenals of each side to between 1700 and 2200 deployed weapons, approximately one-third the levels that were allowed under START I. Thus, this new arms control agreement was one bright spot in what was looking increasingly like a rapid unraveling of the Cold War stand down in the second decade after the end of the Cold War.

Another bright spot was the election of Barack Obama as President of the United States in 2008. Obama came into office after making the worldwide elimination of nuclear weapons one of the major goals of his presidency. He had also been an outspoken opponent of the war in Iraq. He saw that he could potentially accomplish both goals – ending the war in Iraq and eliminating nuclear weapons – if he could accomplish what he called a "reset" of the U.S.-Russia relationship. The timing was auspicious, because both START I and SORT would be expiring during the new President's term. Accordingly, he invested the time, effort, and resources required to engage in negotiations with Russia to draft a new strategic arms control treaty.

Negotiations and drafting of NEW START took place throughout much of 2009. During a trip to Prague on April

8, 2010, President Obama met with Russian President Dmitri Medvedev, and both presidents signed NEW START. This new arms control agreement requires the elimination of half the missile launchers on both sides, and limits total nuclear weapons to 1500 each, a significant cut from the levels of START I. Of even greater significance, NEW START established a new inspection and verification regime to assure treaty compliance.

Debate in the U.S. over NEW START ratification took place throughout the Spring and Summer of 2010, and into the Fall. Many prominent scientists and scientific organizations, such as the American Federation of Scientists, argued strongly in favor of ratification. Almost all leading Democrats were also for ratification. Republicans were split. Some Republicans based their opposition on arguments that Russia could not be trusted to comply with the treaty limits. Others argued that the treaty disadvantaged the U.S. Since ratification needed a two-thirds vote in the Senate it was important to proponents to have strong bipartisan support for treaty ratification.

Republican support for ratification, however, was nearly derailed by a significant new political development that took place in the U.S in 2010. The new development was the hostile takeover of the Republican Party by the TEA Party insurgency. The TEA Party was an invention of Fox radio talk-show host Glenn Beck, who was trying to incite a revolt against paying taxes; the name of the new party comes from Beck's claims that Americans were "taxed enough already," and was meant to liken his supporters to the patriots in colonial Massachusetts who protested British taxes on tea by boarding British merchant vessels in Boston Harbor and throwing overboard crates of imported tea. What started as an erstwhile tax revolt quickly morphed into a large and widespread movement to first take over the Republican party by running TEA Party radicals against establishment Republicans in the 2010 primary elections, and then by winning control of both the House and the Senate from the Democrats in the 2010 midterm elections.

The TEA Party was – and still is – a loose and fluid confederation of libertarians, free-market radicals who opposed government intervention in the operation of the market (so they were generally opposed to taxation, government regulation, and government spending on social welfare programs and health care); an odd assortment of white supremacists, white nationalists, neo-Confederates, anti-government militias, and other far-right fringe groups that felt they were going to be further marginalized by the election of America's first African-American president; and Christian evangelicals and fundamentalists, many of whom felt that a black progressive Democrat was a threat to conservative Christian values.

In 2010 the TEA Party had the perfect issue around which to organize: the passage early that year of President Obama's signature piece of legislation, The Patient Protection and Affordable Care Act (ACA), colloquially known as Obamacare. The ACA was the most comprehensive reform of health care and expansion of health insurance coverage in the U.S. since the creation of Medicaid and Medicare in 1965. Since President Obama came into office with a Democratic majority in both the House and the Senate, he was able to get health care reform done despite nearly unanimous Republican opposition. Tea Party adherents were generally opposed to NEW START ratification. Opposition to the ACA, however, and calls for its rescission, was the one single issue around which the TEA Party could organize and unite all the disparate elements that made up the party.

TEA Party efforts were surprisingly successful. In the November 2010 midterm election, the Republicans captured the majority in the House of Representatives with the largest change in seats from one party to another since 1948. In the Senate, the Democrats lost seats but maintained a slim majority.

President Obama understood that once the composition of

the Congress changed in January 2011 it would be much more difficult to get NEW START ratified. He, therefore, pushed hard for its ratification in the lame-duck session of the Congress after the November election, despite the opposition of some influential Republicans. The key players, however, were Senator John Kerry, the Democratic Chairman of the Senate Foreign Relations Committee, and the Committee's Ranking Member, Republican Senator Richard Lugar. This was the same Richard Lugar who had co-sponsored the 1991 legislation that set up the Cooperative Threat Reduction Program, as discussed in the previous chapter. Lugar's strong support for the treaty was decisive. So was the support of six former Republican Secretaries of State.[10] The U.S. Senate voted 71 to 26 to ratify NEW START on December 22, 2010. The Russian State Duma took its first ratification vote in favor of the treaty two days later. The Duma's final ratification was completed and the treaty became effective on February 5, 2011, superseding SORT.

The pause in the unraveling of the Cold War stand down represented by NEW START ratification was not to last. In President Barack Obama's second term, starting in 2013, the fragile peace in Ukraine between political forces pushing for closer economic ties to Europe and those pushing instead for closer economic ties to Russia came completely undone. It started when the pro-Russian Ukrainian President, Viktor Yanukovych, suspended an evolving association agreement with the European Union. This decision unleashed a wave of protests throughout the country (but mostly centered in the capital, Kiev). The protests turned violent in January 2014, leading over the following months to the overthrow of Yanukovych, the election of a new pro-Europe President, Petro Poroshenko, and the election of a new government. Meanwhile, Russian separatists in the Donetsk region of eastern Ukraine, who were not supportive of the revolution taking place in Kiev, formed their own breakaway republic with support from

Russian troops, unleashing a civil war in Ukraine between the new central government and the separatists in Donetsk.

Russia's responses to the revolution in Ukraine had consequences for the U.S.-Russia relationship that are still being felt to this day. Russian President Vladimir Putin moved first to send Russian troops to take over Crimea, a large peninsula that juts into the Black Sea from the southern coast of Ukraine. Crimea had been, during Soviet times, the base for the Soviet Navy's Black Sea fleet. Under-cover Russian troops (that is, army troops in uniforms that were stripped of identifying insignia), without firing a shot, moved into Crimea and disarmed and neutralized the Ukrainian armed forces there. Then, Russia organized a popular referendum for Crimea's residents to approve Russia's annexation of Crimea. The formal annexation was completed on March 18.

The United Nations,[11] the European Union, and the U.S. all declared the annexation illegal and invalid, and called for restoring the territorial integrity of Ukraine. Not only was Russia diplomatically isolated as a result of their annexation of Crimea; but economic sanctions were also imposed on Russia, most of which are still in effect as of this writing (2020). Meanwhile, the fighting in the Donetsk region continues, with Russia militarily supporting the separatists, and the U.S. supporting the Ukrainian government. U.S. support to Ukraine includes the supply of weapons. Thus, Russia's actions in Ukraine began a renewed and accelerated unraveling of the Cold War stand down.

Then, just as it became difficult to see how things could get worse, something else unexpected happened in 2016 that further accelerated the unraveling of the Cold War stand down: the election of Donald Trump as President of the United States.

The election outcome was unexpected because Trump, the Republican candidate, had not held any elected office previously, had no experience in any aspect of public policy, was widely viewed as singularly unqualified, and was well behind in the polls right up to election day, even though he had wide support among TEA Party adherents. Indeed, he did lose the popular vote to his Democratic rival, Hillary Clinton, who had served as Secretary of State in President Barack Obama's first term. Trump won the election, however, because he eked out a narrow victory in the Electoral College. Pundits and scholars still debate the reasons for Trump's surprise win, but that debate goes well beyond the focus of our story here and thus will not be discussed.

Except for one thing: the role of Russia in the 2016 U.S. election. Russian intelligence agencies ran a broad disinformation campaign using social media platforms in an attempt to damage and discredit Hillary Clinton[12] and sow division and distrust in the American electorate. The fact of Russian interference in the 2016 election is not in dispute (except by President Trump himself). What is in dispute is whether or not President Trump's campaign illegally or improperly cooperated with or coordinated with or aided this disinformation campaign. This question was the subject of probes by committees of Congress, the U.S. Justice Department, and the U.S. intelligence community, during the campaign and after Trump became president. It is beyond dispute, though, that Donald Trump himself cheered on the Russian efforts. His campaign rallies were frequently highlighted by chants of "lock her up," "her" being a reference to Hillary Clinton, and were usually incited by Trump repeating some bit of Russian disinformation about her.

In addition to President Trump's dismissive attitude towards Russian interference in the 2016 election, his ill-advised actions and statements with respect to arms control, NATO, international cooperation, science, the rule of law, and the norms of democratic governance raised alarms. Here the focus is only on what President Trump did with two major arms control treaties that have underpinned the Cold War stand down, INF and NEW START.

The Intermediate-range Nuclear Forces (INF) Treaty, as discussed in Chapter 8, was the arms control treaty signed by U.S. President Ronald Reagan and Soviet Premier Mikhail Gorbachev in 1987, marking the beginning of the end of the Cold War. The treaty was still in force when Donald Trump became President in January 2017. Then, President Trump announced on October 20, 2018, that the U.S. would withdraw from the treaty. On February 1, 2019, the U.S. formally suspended the treaty, starting a six-month clock until treaty withdrawal. In response, Russia also suspended the next day. The U.S. formally withdrew on August 2, 2019.

Many of the same arguments that were made in justifying U.S. withdrawal from the ABM Treaty seventeen years earlier were made again this time. In particular, the U.S. charged that Russia was developing nuclear-capable ground-launched cruise missiles that have a range that is prohibited by the treaty. Russia countered this argument by claiming the treaty covered only ballistic missiles, not cruise missiles, and that expansion of NATO forces into eastern Europe was a threat to them. Instead of engaging the Russians in negotiating a new INF treaty that would take account of these more recent technological and political developments, President Trump decided instead to simply scuttle the treaty.

Every U. S. President since the end of World War II and the beginning of the Cold War has engaged the Russians in arms control negotiations, and several important treaty agreements have been signed, as discussed in this book. All of these treaties have been aimed at avoiding the apocalypse of nuclear war, always present as a possible endpoint of the Cold War conflict. All Presidents, that is, until President Trump.

This fact became a particularly worrisome matter because the newest strategic arms control treaty, NEW START, a signature accomplishment of President Obama, was set to expire in

February 2021. Treaty provisions, however, allowed a five-year extension. Russia had several times indicated their interest in extending this treaty. President Trump claimed, however, that this treaty was one of several "bad deals" President Obama negotiated,[13] yet he made no move to negotiate a better one. This became a worrisome development, because allowing the treaty to expire would have put the world right back into an escalating and unstable nuclear arms race. It became clear that, as the deadline date approached for treaty expiration, it would ultimately be up to who was President in 2021, following the election of 2020, to either let the treaty expire or extend it to buy time to negotiate a new one. The two candidates for President in that election, President Trump standing for re-election, and his Democratic challenger, Joe Biden, were on opposite sides of this important issue. Proponents of treaty extension, including Joe Biden, advanced three reasons why President Trump's judgment could not be trusted on any matter concerning Russia.[14]

The first argument was that President Trump displayed a dangerous disdain of expert advice, and an alarming ignorance of science. According to numerous interviews with former staff published in The Atlantic magazine[15] and in newspapers and, more recently, in a book entitled "Trump and His Generals" by Peter Bergen,[16] he had discontinued regular intelligence, military, and foreign policy briefings and ignored written staff briefings on complex issues like arms control. At the same time, he had taken actions in direct contradiction to expert advice, like withdrawing the U.S. from the Iran nuclear agreement and the Paris Climate Agreement; precipitously withdrew U.S. troops from northern Syria, putting America's Kurdish allies in the fight against the Islamic State terrorist group at risk of annihilation while simultaneously undermining the NATO alliance; and upended the military justice system by pardoning convicted war criminals.

Second, President Trump continued his refusal to

acknowledge that Russian intelligence ran a massive and well-organized disinformation campaign during the 2016 election campaign which played a role in helping him get elected. Inexplicably, he also continued to be dismissive of the threat to peace and security posed by Russia's annexation of Crimea and their armed incursion into Ukraine. It is true that economic sanctions on Russia first imposed during the Obama administration, as discussed above, were tightened during the Trump administration, but this was done in a bipartisan effort in Congress, not because President Trump pushed for it. Indeed, President Trump had no coherent foreign policy strategy with regard to Russia, at least not one that he articulated.

Third, there was President Trump's propensity to act only in his own personal business and/or political interests, or, worse, to conflate his personal interests with the national interest. His using Congressionally approved military assistance to Ukraine as leverage to get the Ukrainian government to announce corruption investigations of Joe Biden's son – an act that sparked his first impeachment[17] – is only one example of his behavior in this regard. Equally alarming to many people were the numerous times he said that people who do not show personal loyalty to him are committing treason.

Any one of these reasons was cause for concern among arms control proponents about President Trump's judgment on any matter concerning Russia. Taken together, these three reasons presented arms control proponents with an unassailable case that his judgment could not be trusted. With respect to issues of nuclear arms control, this put the U. S. in a very dangerous situation.

Joe Biden won the 2020 election, was inaugurated as President on January 20, 2021, and shortly thereafter did sign a five-year extension of NEW START. So, the immediate threat to further unraveling of the arms control regime was averted.

Nonetheless, as we approach the middle of the third decade

of the 21st century, the world still seems to be on a course to re-ignite a new Cold War and raise again the specter of nuclear apocalypse. In addition, other apocalyptic threats to human existence are coming to the fore, including global disease pandemics and catastrophic weather events triggered by global climate change. Yet, the U.S. seems ill-equipped to deal effectively with these threats. The question that arises is this: what lessons have Americans not learned about how the Cold War ended?

1. Steven Kotkin argues persuasively in *Armageddon Averted: The Soviet Collapse, 1970-2000* (Oxford University Press, 2001) that the very fact so many Soviet government bureaucrats simply stayed on the job in the new Russian government was a large factor in preventing the sudden collapse of the Soviet Union from devolving into apocalyptic nuclear war. The fact that the newly independent states of Kazakhstan, Belarus, and Ukraine agreed to give up the nuclear weapons they had inherited from the Soviet Union (or transfer them to Russia) was also helpful.
2. Tajik is an Indo-European language, in the same language sub-family as Persian (Farsi). So is Kirgiz. On the other hand, Uzbek, Kazakh, and Turkmen are all distinct Turkic languages. The "lingua franca" of the five formerly Soviet Stans of Central Asia is Russian.
3. Steven Kotkin presents a more dystopian view of these post-Soviet Central Asian Stans in "Trashcanistan," *The New Republic*, April 15, 2002.
4. Frederick Kunkle, *Washington Post*, August 18, 2008.
5. Condoleezza Rice in the *Washington Post*, August 8, 2018.
6. The original member countries of NATO in 1949 were Belgium, Canada, Denmark, France, Iceland, Italy, Luxembourg, Netherlands, Norway, Portugal, United Kingdom, and the United States.

7. The expansion of NATO into former Warsaw Pact countries and former Soviet Republics has been very controversial even within NATO and among U.S. officials. In addition, there is not even agreement on what has and has not already been agreed. For example, there are ongoing debates among scholars, historians, and numerous public officials about whether or not there were any informal understandings stemming from the first post-Cold-War NATO expansion that took place with German re-unification in 1990; Gorbachev himself has said contradictory things about this at different times, claiming in his *Memoirs* (Doubleday, London, 1996) that there was such an understanding, then saying in a later report that the issue was not even discussed then.
8. The other four Yugoslav republics that have joined NATO are: Slovenia in 2004; Croatia in 2009; Montenegro in 2017; and North Macedonia in 2020.
9. These changes to the U.S. nuclear force structure had less to do with START II compliance than with carrying through with long-planned modernizations. Even before START II the U.S. had planned to replace the Peacekeeper ICBMs with fewer but more modern and more capable Minuteman III missiles. It was also part of the long-range plan to strengthen the submarine-launched ballistic missile arm of the triad.
10. Opinion piece by Henry A. Kissinger, George P. Shultz, James A. Baker III, Laurence S. Eagleburger and Colin L. Powell published in the *Washington Post*, December 2, 2010; see also opinion piece by Condoleezza Rice, *Wall Street Journal*, December 7, 2010.
11. United Nations General Assembly Resolution 68/262, entitled "Territorial integrity of Ukraine," adopted on March 27, 2014, with 100 nations voting for, 11 against, 58 abstentions and 24 absent.

12. Any serious discussion of what motivated the Russian disinformation campaign is mostly speculative. No one truly knows whether it was that Russia really did not want Hillary Clinton as President and for what reason, or really wanted Donald Trump as President and for what reason, or simply wanted to discredit democratic elections in the eyes of the Russian populace. It could have been some combination of all three motivations, or a different motivation altogether. Ultimately, it does not matter: the effect in the U.S. was the same.
13. *Reuters*, 9 February 2017.
14. The arguments presented here on President Trump and Russia were first published by the author as an opinion piece in the *Gettysburg Times*, March 30, 2020.
15. See, for example, the article by Eliot Cohen in the October 2017 issue of *The Atlantic* and the one by Jeffrey Goldberg in the October 2019 issue.
16. Peter Bergen, "Trump and His Generals," Penguin Press, 2019.
17. Rep. Eric Swalwell, *Endgame: Inside the Impeachment of Donald J. Trump*, Abrams Press, 2020.

Chapter 11

Lessons Not Learned

It was the imprisonment and exile of the three principal founders of the Moscow Helsinki Watch Group – computer scientist Anatoly Shcharansky, physicist Yuri Orlov, and physicist Andrei Sakharov – that prompted the worldwide scientists' boycott of the Soviet Union.

Anatoly Shcharansky was born on January 20, 1948, in the Ukrainian city of Donetsk. After graduating from Moscow Physical-Technical Institute in 1972 with the equivalent of a Master's degree in mathematics and computer science, he took a job with a Soviet oil and gas company. In 1973, after he applied for a visa to emigrate to Israel, he joined the growing ranks of "refuseniks." His visa application was denied on the basis of "state security," even though neither his thesis work at the university nor his work at the oil and gas company was connected to classified or military research. No matter, because, as we have already seen in Chapter 3, information was so tightly controlled in the Soviet state that any information that was not officially released by the state, whatever its subject, was deemed "classified." Shcharansky at first suffered the usual fate of refuseniks: he was fired from his job. He soon became an activist in the Jewish emigration movement, rising rapidly to a leadership position. Leadership in the Jewish emigration movement meant that he collected and gave information to the foreign press in Moscow about the plight of refuseniks. Such activity, of course, marked him as a "spy" in the eyes of the state security apparatus in the Soviet Union. On March 15, 1977, Shcharansky was jailed after charges of being a spy for the American CIA appeared in the Soviet press. He was held for over a year in isolation in the KGB prison in Moscow,

Lefortovo Prison, while formal charges were being prepared against him. At his show trial in July 1978, he was accused, among other things, of espionage as a CIA agent, a charge that then-President Jimmy Carter vigorously and repeatedly denied. Shcharansky was sentenced to thirteen years of prison and prison camp, strict regime, on July 14, 1978. Strict regime meant that he would be subjected to hard labor, poor food and living conditions, and inadequate – even non-existent – medical care. During the Stalin years, many people sentenced to strict regime labor camp did not survive the experience.

Shcharansky's closing words to the court that sentenced him are contained in Appendix IV.

Shcharansky was finally freed as the result of an East-West prisoner exchange. In February 1986, eight years into his thirteen-year sentence, Anatoly Shcharansky was escorted across the Glieneke Bridge into West Berlin, a free man. He then went to Israel, rejoining his wife, Avital, who had received permission to emigrate in 1974. He had not seen her in 12 years. After his arrival in Israel Shcharansky became active in Israeli politics, holding down several government posts, and was from 1996 to 2003 the leader of the Russian immigrant party in Israel.

Yuri Orlov, unlike Shcharansky, was of an earlier generation and was neither Jewish nor involved in the Jewish emigration movement. In fact, he started his career as a member of the Communist Party, having served in the Soviet army in World War II, and then studied physics first at Moscow State University and then at the renowned Institute of Theoretical and Experimental Physics in Moscow. His outspoken support of democratic reform in the Soviet Union, however, got him into continual trouble with the authorities. After speaking out at a Communist Party meeting in Moscow in 1956 he was expelled from the Party and removed from his position at the Institute. Orlov then relocated to Yerevan, Armenia, where he spent the next sixteen years building his career and his reputation in physics. He earned a

Candidate of Science degree (equivalent to the Ph.D. degree in the United States) in 1958, a Doctor of Science degree in 1963, and was elected to the Armenian Academy of Sciences. During these years he developed an expertise in high-energy particle accelerator physics. He published many scientific papers, which brought him to the attention of scientific colleagues in the West. In 1972 he returned to Moscow to work at the Institute of Terrestrial Magnetism and Radio-wave Propagation, and, of course, found himself again in trouble with the authorities because of his human rights activities in the capital. In May 1976 Orlov became the first Chairman of the Moscow Helsinki Watch Group, co-founding the Group with Shcharansky and Sakharov. The following February he was arrested, and after a trial in May 1978 on charges of "anti-Soviet activity," he was sentenced to seven years in prison followed by five years of internal exile.

At the beginning of October 1986, less than eight months after Shcharansky went free and eight years into his 12-year sentence, Yuri Orlov was released from Siberian exile. Just two weeks later he appeared at a meeting of the Committee on Human Rights of the U.S. National Academy of Sciences in Washington, D.C. He expressed his view that the new Soviet leadership was invested in "broadening its scientific contacts. For the first time, Soviet leaders have truly realized the great depth, the great degree, to which they are lagging behind." He also attributed his good fortune in getting released to the attention given to his case by Western scientists. He was quite clear in his belief that the actions of Western scientists in raising specific cases with Soviet scientists, the boycott, the termination of exchange programs – all had an effect in preventing further persecutions and harassment. Orlov is now an Emeritus Professor of Physics at Cornell University in the United States.

Even though the three scientists who were the focus of the scientists' boycott – Shcharansky, Orlov, and Sakharov – were all released in 1986, this did not mean that the overall situation was fundamentally changed. Indeed, in many ways, the human rights situation was as grim as ever. While Shcharansky, Orlov, and Sakharov went free, forty other Helsinki Watch Group members remained in prison. Suppression of Jewish cultural activity was intensified, particularly with the arrests of numerous Hebrew teachers. Hebrew was not an officially recognized language, publication or import of books in Hebrew was banned, and teaching it was a proscribed activity. One such teacher, Alexey Magarik, was arrested on trumped-up charges of drug possession. Indeed, about half the Jewish prisoners in the Soviet Union at the end of 1986 had been arrested or put on trial since Gorbachev came to power. In addition, emigration remained dismally low; after the 1979 high of more than 51,000 émigrés, the numbers plummeted to less than a thousand a year by 1984, and remained low for the rest of the decade (see Appendix II).

Thus, at the end of 1986, the only thing that seemed constant about the human rights situation in the Soviet Union was its inconstancy. While some people were released and others were able to emigrate with no hassle, still others were persecuted and prosecuted, a situation very similar to that in 1978 when Shcharansky and Orlov first went to prison. In that year two American journalists, Harold Piper of the *Baltimore Sun* and Craig Whitney of the *New York Times*, were convicted of libel by a Soviet court and ordered to pay a fine. An American businessman was given a five-year suspended sentence for violating currency laws, a charge he denied. On the other hand, several long-term refuseniks were given permission to emigrate after a personal intervention by Senator Edward Kennedy of Massachusetts. These included Benyamin Levich, a prominent Soviet scientist who had first applied for an exit visa in 1972. So, despite the resolution of the high-profile cases of Shcharansky and Orlov by mid-October 1986, the fundamental situation looked little different from 1978. Gorbachev, however, had only been in power for about a year and a half at this time, and the *perestroika* reforms were only just getting underway.

The third person who was the focus of the scientists' boycott, Andrei Sakharov, as we have already seen in Chapter 8, was released from internal exile in December 1986 by Mikhail Gorbachev, triggering a sequence of events that led directly to the end of the Cold War. Sakharov, unfortunately, did not live to see all these things. In November 1988 he made his first-ever trip abroad, coming to the United States, where he met President George H. W. Bush, and to Western Europe, where he met with several European heads of state. In March 1989 he was elected to the Congress of People's Deputies, where he worked to organize a reform movement within the Soviet government. In late 1989 he was planning his third trip abroad since his release from internal exile. He had a visiting appointment awaiting him at the University of California, Berkeley, and had several scheduled speaking engagements, including a scientific seminar at the Los Alamos National Laboratory at the author's invitation, when he finally succumbed to the ill health that was the legacy of his years of maltreatment during his hunger strikes and internal exile in Gorky. Andrei Sakharov died of a heart attack in Moscow on December 14, 1989.

As for the other principal player in the events of the late 1980s and early 1990s that saw the ending of the Cold War, Mikhail Gorbachev became the first elected President of the U.S.S.R. in 1990, the same year he was awarded the Nobel Peace Prize. In August 1991 Russian communists staged an attempted coup to wrest control of the government from Gorbachev and undo the democratic reforms he had put in place. Gorbachev was actually kidnapped by the coup plotters during a vacation trip to the Crimea, and was freed three days later after Boris Yeltsin, then Mayor of Moscow, and his democratic supporters in Moscow faced down the tanks in the Moscow streets and foiled the coup attempt. Three days later Gorbachev resigned his position as head of the Communist Party. Then on Christmas Day, he resigned as President of the U.S.S.R. and the Soviet Union was dissolved as a country. In an astonishingly short period of time, the Communist Party fell from power in the Soviet Union, and the Soviet Union itself ceased to exist.

Boris Yeltsin became the first elected President of a new democratic Russian Federation, but Gorbachev tried to stage a political comeback in 1996, running as a candidate in the presidential election that year. The voters of Russia, however, decisively rejected the man who had guided them out of the rigid control of the Communist Party. The Communist Party candidate, Gennady Zyuganov, won less than 25% of the vote, a resounding defeat for communism at the polls, but Gorbachev received less than 1% of the vote. Boris Yeltsin was returned to the Presidency, even though his leadership qualities were called into serious question over the previous five years of political and economic turmoil and upheaval in Russia.

With the election of Vladimir Putin to succeed Boris Yeltsin as President in March 2000, Russia began its long slide into right-wing nationalist and populist authoritarianism. The dissolution of the Soviet Union, as discussed in the last chapter, unleashed social and ethnic tensions that had been held in check by the Soviet police state. Some of these tensions erupted into open warfare, as in the Russian province of Chechnya and the former Soviet republics of Georgia and Ukraine. Antisemitism, always endemic within the Russian population, and exploited shamelessly by the Soviet state during the Cold War, has not disappeared in the new Russia. Right-wing nationalist parties continue to exploit antisemitic sentiments, and to work against cooperation and accommodation with the West. The new Russian democracy was also plagued by corruption, an explosion of organized crime, and "crony capitalism," a situation in which a few friends of the people in power make vast sums of money by exploiting their political connections.

Given all these circumstances it is perhaps not unexpected that a right-wing demagogue like Putin rose to power in Russia and has reversed progress towards democratization. Compounding the problem, with Sakharov dead and Gorbachev in political exile, there are really no intellectual leaders left to guide the transition to real democracy in Russia. Most of the people who can provide this guidance have left the country, including many of Russia's best scientists. It is not at all clear that the Russians who have remained have learned the right lesson from how the Cold War was brought to an end.

Perhaps an even bigger risk for the United States is that it is not at all clear that Americans have learned the right lesson from how the Cold War ended. The risk for the United States is that Americans, just like the Spanish in sixteenth-century Peru, have learned the wrong lesson from their "victory" in the Cold War. Sixteenth-century Spaniards interpreted their easy victory over the Inca Empire as proof of the superiority of their Catholic faith and European culture, so they did not bother to explore the weaknesses in Inca thinking that they in fact shared with the Incas, particularly the concept of "divine right" and central control of information that flowed from this idea. In the wake of the Spanish conquest, everything changed for the Spanish and for the New World peoples except their fundamental way of thinking. This left the Spanish Empire itself vulnerable to weakening and collapse because of their adherence to wrong notions and a misreading of history.

Similarly, the lesson propagated by the American right is that American moral or ideological superiority – what is often framed as "American exceptionalism" – is what won the Cold War, and that this superiority must be maintained with military strength and power. It was this attitude on the part of the American right that led in 2000 to the U.S. Senate's rejection of CTBT ratification. President Bill Clinton had originally sent the Treaty to the Senate for their advice and consent to ratification, as required by the U.S. constitution, in September 1997, one year after being the first head of state to sign the Treaty. The CTBT, as explained in Chapter 9, was, along with the Stockpile Stewardship Program, a "grand compromise" between the American left and the right, a compromise that was intended to facilitate the final ending of the Cold War and a new beginning in the great-power relationship between Russia and the United States. The timing, however, was terrible, because the President was caught in a highly partisan battle over impeachment, which deepened the partisan divisions in the U.S. Congress and among the American people. The Republican-led U.S. Congress was in no mood to accommodate the Democratic President's wishes on the CTBT, especially when several leading Republican conservatives, particularly Senator Jesse Helms, Chairman of the Senate Foreign Relations Committee, saw the CTBT as putting the United States at a great military disadvantage. Helms even blocked hearings on the Treaty, and the President was in no position to push for it. Nonetheless, during the President's State of the Union Address on January 19, 2000, after he survived the impeachment vote in the Senate six months previously, he called upon the Senate to ratify the Treaty, asking the Senators to "approve the Treaty now to make it harder for other nations to develop nuclear arms, and to make sure we can end nuclear testing forever."

President Clinton also urged the Russian Duma (parliament) to do likewise. On June 30, 2000, the Duma voted to ratify the CTBT, leaving only China and the United States as the declared nuclear powers having signed but not ratified the Treaty as of that date. Yet, the President still did not expend any political capital – probably because he did not have any to spend with this largely hostile Congress – on making the case for ratification and pushing for hearings and a ratification vote. When it became clear that there were not enough votes to ratify, the Republican Senate Majority Leader, Senator Trent

Lott, scheduled a ratification vote over the objections of the President, and the CTBT became the first arms control treaty in history to be turned down for ratification, and in a largely party-line vote.

The Senate's rejection of the CTBT threatened to unravel the "grand compromise" on which the end of the Cold War was being built. The ideological split in the United States has thus not closed with the end of the Cold War, but has instead continued and perhaps even widened, and threatens to undo the efforts to bring about a final conclusion to the Cold War ideological conflict and its attendant military confrontation.

Despite the great advances in science, scientific thinking and scientific method, and the general intellectual enlightenment of the wider population, what is most remarkable since the scientific revolution is the stubborn persistence of irrationalism. This irrationalism is manifested in many forms, including religious intolerance and zealotry, racism, nationalism, elitism, and authoritarianism. All of these irrational ideologies are based on "revealed" truths, not scientifically established truths. These revelations usually come to the true believer from someone recognized by the believer as a higher authority. The greater certainty the believer has in the truth of his or her belief the greater passion with which the belief is held, even in the face of overwhelming evidence that the belief is false. In fact, a great many people admire this quality of passionate belief in others. How many times do we hear that a particular person is desirable as a leader because he or she stands by the "courage of his (her) convictions"? Never mind that the convictions may be wrong; it is the passion that is valued. We tend to dismiss the person who approaches issues with skepticism and incertitude.

Zealotry is fueled by belief and passion, not reason, and the believer almost always assumes his or her superiority to those who do not hold to the same beliefs. Zealotry combined with political power presents a particular danger for democracy.

This is because the zealots, motivated by the belief that they are the sole possessors of truth, must maintain a hierarchical power structure and control of information. Even those people without power who share the beliefs of the people with the power come to support the power structure because they believe that the structure maintains what is right and true, and that those people with the power earned it on merit. Even scientists, rational beings that we are, fall into the same trap. Almost all large scientific laboratories and institutions are hierarchically structured, with the top people maintaining their power over the working lives of the others by propagating the belief that the organization is a meritocracy and that they earned their right to be at the top of it because of the superiority of their ideas or work.

As long as the purpose of political organization is viewed as a hierarchical structure to maintain the power of one ideology or another, the zealots or ideologues will always win, at the expense of true democracy. If the ideologues are the ones out of power they will feel grievously alienated and fight more passionately to obtain it. If they are in power, they will likewise fight more passionately to retain it.

An example of this phenomenon was the final outcome of the U.S. Presidential election of 2000, in which the candidate who lost the nationwide popular vote was ultimately declared the winner with the help of the passionate zealotry of his ideological supporters. Without the 25 electoral votes of the State of Florida neither the Democratic candidate, Al Gore, or the Republican candidate, George W. Bush, had won enough electoral votes to win the election, and the popular vote count in Florida was, to within the margin of uncertainty because of uncounted and under-counted ballots, a statistical tie. The Republican candidate ultimately was awarded the Florida electoral votes mainly because the ideologues were on his side in the ensuing post-election legal battles, along with the state's political power structure, including the Governor (who was the candidate's brother), the State Legislature, and the Secretary of State (who was the state official in charge of running the election and simultaneously the candidate's state campaign manager, a clear conflict of roles). Even though the Republicans raised reasonable questions concerning what uncounted and under-counted ballots to count or recount, and how it was to be done, the Republican position from the start, since the Republican candidate was slightly ahead in the vote count as of election night, was to stop vote counting by any and all means. This included, in addition to the endless legal challenges to all legal recount requests by the Democrats, numerous mass demonstrations, the propagation of the false belief that the Democrats were trying to "steal" the election, and the outright intimidation of official vote counters to stop the count.

The ideologues prevailed in the 2000 election because they had a lot at stake. The CTBT was only one of many issues that motivated the passions of the Republican ideologues, albeit a relatively minor issue in the 2000 campaign. The two candidates, however, had very different positions. Al Gore, like President Clinton, was a strong supporter of the CTBT. George W. Bush held to the same position as most of his party, opposed to the CTBT. As explained in Chapter 9, the CTBT is an important component in the endgame strategy of the Cold War. One is compelled to conclude that Americans – at least those on the right and those who support them – have not learned the right lesson from the end of the Cold War.

So, what is the right lesson? The right lesson is the lesson that Andrei Sakharov spent his life teaching us. It is that there must be a new linkage between science and democracy, strengthening both. To explain what I mean by this I must first explain what is meant by science in this context, and how it is linked to democracy. First, however, it would be helpful to say what science is not.

Science does not include any knowledge that is obtained by belief and which cannot be either verified or falsified by direct observation or experimentation. This includes UFOlogy (a belief that the Earth has been visited, and continues to be visited, by intelligent aliens from other planets), parapsychology, psychic surgery and spiritual healing, demonology and witchcraft, astrology, and creationism. Many, if not all, of these fields of endeavor, freely borrow the language of science, and many of their practitioners and believers even claim to use the methods of science to arrive at their "truths."[1, 2] Such claims, however, are wildly false, as we shall see.

There are many believers, however, who do not pretend to justify their beliefs on the basis of science, but instead claim that "other ways of knowing" what is true are just as valid, and perhaps more so, than the scientific method. Indeed, beliefs have often served throughout human history to guide behavior to the benefit of the individual believer, and sometimes also to the benefit of the wider society. Religiously based codes of ethics have been, and still are, used to guide behavior in beneficial directions. Thus, one can ask what is wrong with someone's belief, for example, in the predictive power of astrological forecasts when the believer may find some beneficial help in everyday decision-making by consulting such forecasts? Or, to take a simpler example: if a person finds, after a number of tries to obtain what he wants, that he succeeds after changing the tilt of his hat on his head, what is wrong with this person holding to the belief that he owes his success to his hat tilt and thus develops an everyday hat ritual? After all, if it works for him what is the harm?

It is precisely this belief in the power of beliefs, however, that is so dangerous, because once the belief fails – for example, an astrological forecast one day turns out to be wrong, with disastrous consequences, as will certainly happen sooner or later – the believer is devastated by a feeling of powerlessness.

After all, if one cannot trust one's own beliefs, what can one trust? But this is precisely the point: "other ways of knowing" often lead to false knowledge in which we cannot, and should not, place our trust. There is also a danger to the wider society. When the believer in the hat ritual, for example, comes to believe that his way of wearing a hat is the only true, right way, and finds himself in a position of power to impose his way on everyone else, then that society is no better off than one ruled by the authority of "divine right."

This belief in the power of beliefs, though, makes it so difficult to undo the mischief that these beliefs cause in our public and community life. It is also why the same ideological battles over the same issues re-arise over and over again in different forms and different venues. Believers in creationism, for example, continue to press their beliefs in divine creation as scientific "fact," while maintaining that biological evolution is false. Just as in the Tennessee trial of the teacher John Scopes in 1929, the creationists still occasionally win some legal or political victories in their passionate efforts to advance their ideology, to the detriment of the greater society. One such more recent example was the election of creationists to the Kansas State School Board, and their success in 1999 in getting the Board to agree to remove all references to evolution from the state-wide science teaching curriculum. The American Physical Society issued a statement in January 2000 denouncing the Kansas School Board decision as "a giant step backward," one that "should sound the alarm for every parent, teacher, and student in the United States." The APS statement goes on to say that:

> *on the eve of the new millennium, at a time when our nation's welfare increasingly depends on science and technology, it has never been more important for all Americans to understand the basic ideas of modern science.*
>
> *Biological and physical evolution are central to the modern*

scientific conception of the Universe. There is overwhelming geological and physical evidence that the Earth and Universe are billions of years old and have developed substantially since their origins. Evolution is also a foundation upon which virtually all modern biology rests.

This unfortunate decision will deprive many Kansas students of the opportunity to learn some of the central concepts of modern science.[3]

The same issue of the American Physical Society News that contains this statement about the Kansas School Board decision also contains the following "What is Science" statement:

Science extends and enriches our lives, expands our imagination and liberates us from the bonds of ignorance and superstition. The American Physical Society affirms the precepts of modern science that are responsible for its success.

Science is the systematic enterprise of gathering knowledge about the universe and organizing and condensing that knowledge into testable laws and theories.

The success and credibility of science are anchored in the willingness of scientists to:

(1) Expose their ideas and results to independent testing and replication by other scientists. This requires the complete and open exchange of data, procedures, and materials.

(2) Abandon or modify accepted conclusions when confronted with more complete or reliable or observational evidence.

Adherence to these principles provides a mechanism for self-correction that is the foundation of the credibility of science.

In other words, science is not simply a collection of facts. It is a *process* for discovering which facts are true and which are not. The process is one of skeptical inquiry, a testing of falsifiable hypotheses. If a claimed result or truth or conclusion cannot be obtained by an independent investigation conducted under identical conditions then it is not scientific truth. Thus, for a "fact" to be accepted as scientific truth it must be both *reproducible and verifiable*. "Revealed" truths and "knowledge" gained from studies of UFOlogy, parapsychology, psychic surgery and spiritual healing, demonology and witchcraft, astrology, and creationism are neither reproducible nor verifiable, and hence not scientific.

Note also that there is an emphasis on *skeptical* inquiry. This means that there is no scientific fact that can be asserted to be true with absolute certainty. A true scientist must always be ready to abandon a "truth," no matter how long it has been held to be true, if new scientifically obtained evidence shows it to be false. This concept is very difficult for non-scientists to understand or accept. Many people mistakenly interpret scientists' lack of certainty and their seemingly endless arguments over the meaning of particular data or observations or theoretical results as casting so much doubt on established scientific truths that these truths can just as easily be false. Thus, creationists, for example, hear scientists debate the evidence for punctuated biological evolution versus continuous evolution and misinterpret these debates as showing that biological evolution is "merely" a theory that is no more likely to be true than the "theory" of creationism. In fact, the debates are only over details of the mechanisms and the timings of evolutionary changes. The scientific evidence for the *fact* of evolutionary changes in biological organisms is so overwhelming that no reputable scientist claims that Darwin's basic idea is wrong. Nor would any reputable scientist claim that it is true with absolute certainty. The *hypothesis* that it is true, however, has withstood a large range of reproducible and verifiable tests, and hence the theory of biological evolution can be accepted as scientifically true with a high degree of certainty. Science does not need absolute certainty to advance understanding. Scientists are perfectly comfortable operating in a world in which only ever-higher degrees of certainty can be attained, but never absolute certainty.

This high degree of certainty in scientific truths comes about only after a long process of open and free debate and exchange of ideas. The whole process of skeptical inquiry, therefore, that forms the basis of the scientific method depends integrally on democratic processes. For scientific truth to be established everyone must be free to participate in the information exchange and the conduct of the tests and experiments and debates. Without such an open, free, and democratic process, it is not possible for a hypothesis to become verified and a result to be reproduced, and hence for the hypothesis to enter the realm of established scientific truth. Indeed, scientific understanding cannot advance when information is controlled by a few recognized authorities and not made available to everyone. This is the fundamental truth about science that Sakharov and other Russian scientists recognized and understood, and what brought them into conflict with a Soviet political system that based its power on the central control of information.

Sakharov, however, recognized much more than science's dependence on democracy. He also recognized that the linkage went in the other direction. Not only is science strengthened by its linkage with democracy, but democracy is also strengthened by its necessary linkage to science. The scientific method can provide a framework for the public policy debates on which the strength of democracy depends. This is precisely the linkage of which Sakharov spoke.

This linkage is also an expansion of the concept of "natural law" that Thomas Jefferson wrote into the Declaration of Independence. Natural law is made in a process that is open and free for everyone to participate, just like science, with the objective, also like science, of arriving at what is verifiably true. Only what is verifiably true can be "natural." Only what is natural can be trusted for the long term to serve the public interest.

Hence, democracy is strengthened by replacing ideology with rationality, by replacing absolute certainty in our beliefs with the high degree of certainty in what we know to be verifiably and reproducibly true as the foundation of our civic life. Democracy is strengthened by applying the same principles of the scientific method that scientists use to arrive at an understanding of the laws of nature to the formulation of the laws that govern the conduct of our public life.

This does not mean that all citizens must become scientists. It does mean, however, that citizens should have a basic understanding of what is involved in the enterprise of science, and a basic knowledge of the principal concepts of science. Scientists have a special responsibility to facilitate this process, as Sakharov has said. Their special knowledge and understanding obligates scientists to play a more central role in the democratic process, and to communicate their knowledge and understanding, insofar as it affects public policy decisions, to the general public. The press has a special and important role to play in this process, too. They can do this in many ways, but this is the subject of another story for another time.

As for the current story, the final ending is yet to be written. As a result of the belief in the power of beliefs, it is very difficult for people to learn the right lessons from historical experiences like the ending of the Cold War. As we have seen, the American right clings to their notion that America "won" the Cold War as a result of the superiority of American military strength and ideology. They continue to propagate the false notion that President Reagan's military buildup and the Star Wars missile defense program were the cause of the collapse of the Soviet Union. The American left, meanwhile, continues to propagate an equally false notion that the moral superiority of their peace and disarmament campaign is what was responsible for the end of the Cold War.

The fact is that as long as people adhere to these false notions the ideological conflict continues. The Cold War has not yet come to a final conclusion. It will finally end only when we all learn the lesson of linkage that Andrei Sakharov taught.

1. *The Demon-Haunted World* by Carl Sagan (Ballantine Books, New York, 1996).
2. *Voodoo Science* by Robert Park (Oxford University Press, Oxford, U. K., 2000).
3. The creationists failed to win re-election in the 2000 election cycle, and in February 2001 the new Kansas State School Board rescinded the ban on the teaching of evolution in the State of Kansas.

Appendix I

Letter from 72 Soviet Academicians Attacking Sakharov

The following letter was published in the Soviet newspaper *Izvestia* (News) on October 26, 1975.

Soviet scientists, as well as the peace-loving community all over the world, are deeply satisfied with the positive development of international life toward the relaxation of international tension and strengthening of peace. With hope in the future we welcomed the results of the Helsinki Conference on Security and Cooperation in Europe as an important step on the road to peace all over the world. Soviet scientists, together with progressive-minded scientists all over the world, have always supported peace, friendship and cooperation among peoples. We fully share and support the peaceful policy of the Soviet Union. Therefore, we cannot help but express our bewilderment and indignation in connection with the decision of the Nobel Committee of the Norwegian Storting to award the Peace Prize to Academician Sakharov whose activity is directed at undermining peace, peaceful relationships of equal rights among different countries, and aimed at arousing distrust among peoples. People of good will on Earth know that the U.S.S.R. consistently pursues a policy of peace and relaxation of international tension, that it was the Soviet government that initiated and persistently calls for prohibition of the testing of atomic weapons, control of the armaments race, reduction of armaments and armed forces, for observation of the principles of respect for sovereignty and non-interference in the internal affairs of other countries, and rejection of the application or threat of application of force. However, Sakharov fights against this policy, calls on the West not to trust the Soviet state, to follow a "hard" line in relationships with the Soviet state, to demand a "fee" for détente, the rejection of the fundamental achievements of the Soviet regime, and essentially to give the opportunity for a free development of capitalism in our country. He speaks about the danger of détente, and together with anti-Soviet persons in the West he frightens people with the threat of war as if this threat comes from our country. Sakharov is always with those whose aggressive activities have many times stretched international tension to the breaking point. He blamed American military circles not for their aggression in South Vietnam and Cambodia, but for their "insufficient decisiveness and consistency" in the realization of that aggression. The freedom and peace which were hard won by Indochinese patriots he called a "tragedy." He blamed countries which support the just cause of the Arab peoples fighting against Israeli aggression.

Declaring himself a defender of humanism and human rights Sakharov expressed his hope that "the Pinochet regime will open an epoch of rebirth and consolidation in Chile." He was "shaken" by the fate of "poor Hess" – the close accomplice of Hitler who was convicted for his fascist crimes against mankind by the International Tribunal. But the Nobel Committee has proclaimed Sakharov as "a voice of conscience of all mankind."

Under the pretext of the struggle for human rights Sakharov comes out as the opponent of the Soviet peaceful policy, of our socialist society. He slanders the great political, economic, social, and cultural achievements of the Soviet people. We are therefore not surprised by the fuss about this prize that has been raised in the West on the part of certain circles which are interested in destroying the relaxation of international tension and the revival of the Cold War and who seek a pretext for vilifying by any means the noble goals and the sincerity of Soviet foreign policy which has attained general recognition and popularity throughout the world.

For true supporters of peace, the decision of the Nobel Committee is unacceptable, and contradicts in its essence the spirit and letter of the basic statutes about this prize. Soviet scientists consider that the award of the Nobel Peace Prize to Academician Sakharov is of an unworthy and provocative nature and is a blasphemy against the noble ideas cherished by us all of humanism, peace, justice, and friendship between peoples and countries.

Signed by the following Academicians:

G. B. Abdulaev, G. A. Avsiuk, A. P. Aleksandrov, V. A. Ambartsumian, M. S. Asimov, A. A. Bayev, N. G. Basov, N. V. Belov, N. A. Borisevich, A. E. Brownstein, A. P. Vanichev, I. N. Vekua, E. P. Velikhov, A. P. Vinogradov, S. M. Volfkovich, C. V. Vonsovsky, B. M. Vul, Ya. S. Grosul, N. P. Dubinin, N. M. Zhavaronkov, Yu. A. Zhdanov, A. A. Imshenetsky, A. Yu. Ishlinsky, A. P. Kapitsa, K. K. Karakeev, M. V. Keldish, F. V. Konstantinov, V. A. Kotelnikov, E. M. Kreps, A. M. Kunaev, G. V. Kurdyunov, A. L. Kursanov, M. A. Lavrentiev, L. M. Leonov, A. A. Logunov, A. K. Malmeister, M. A. Markov, G. I. Marchuk, Yu. Yu. Matulis, N. V. Melnikov, I. I. Mints, E. N. Mishustin, A. N. Necmeyanov, A. I. Oparin, B. E. Paton, B. N. Petrov, N. A. Piliugin, B. B. Piotrovsky, P. N. Pospelov, A. M. Prokhorov, O. A. Reautov, A. M. Rumyantzev, K. M. Ryzhikov, B. A. Rybakov, A. S. Sadykov, N. N. Semenov, D. V. Skobeltzin, G. K. Skriabin, V. M. Smiznov, V. I. Spitzin, V. D. TImikov, A. N. Tukhanov, A. A. Trofimchuk, V. M. Tuchkevich, P. N. Fedoseyev, N. P. Federenko, G. N. Flerov, A. V. Fokin, A, N, Frumkin, M. B. Khrapchenko, N. V. Tsitsin, V. A. Englehardt

Appendix II

Jewish Emigration from U.S.S.R., 1969-1989

There was always a very small trickle of Jewish emigration from the Soviet Union to Israel, mainly elderly Jews who were reuniting with family members who were already there. The Soviet Union, in fact, was the first country to accord diplomatic recognition to the new state of Israel at its founding in 1948. The situation changed dramatically, however, after the 1967 Six-Day War in the Middle East. The Soviet Union severed diplomatic relations with Israel then, and even the tiny trickle of emigration that had been allowed up to that time was stopped. More important, there was a great awakening of Jewish consciousness in the Soviet Union. This awakening had already begun as far back as the Khrushchev "thaw" of 1956 as a separate but integral part of the overall beginning of the Soviet dissident movement, but the Six-Day War was a great stimulus, and an exploding demand for emigration its result. Even by 1969, the Soviet government could no longer contain the growing demands for emigration, and the exodus began. Only emigration to Israel was allowed, and only for purposes of "family reunification." Since the Soviet Union did not have diplomatic relations with Israel, and there were no direct flights between the two countries, all émigrés went either to Vienna or to Rome, and then were rerouted by a Jewish aid organization. A growing percentage of émigrés, once arriving in Vienna or Rome, opted to go to the United States or another Western country instead of to Israel. The total number of émigrés each year is shown in the graph below.

After the effective end of the Cold War in 1989, there was an explosion of emigration, with most people going to Israel. From the beginning of the present wave of emigration at the

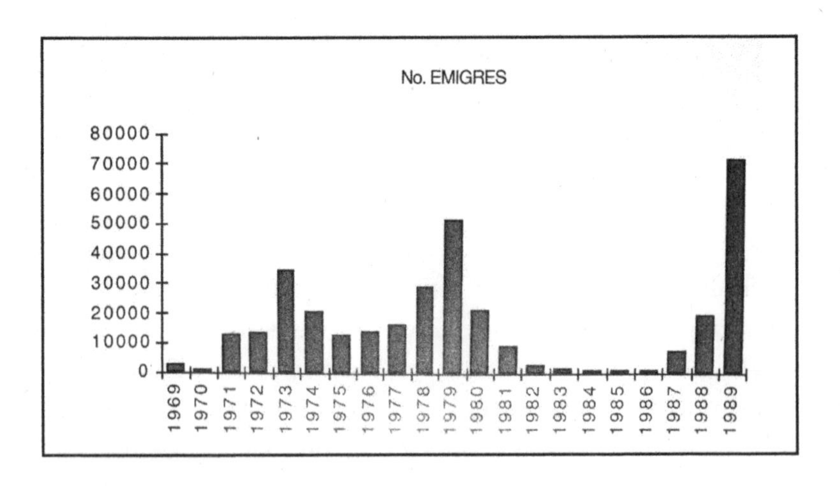

end of 1989 through 1998, 769,650 people emigrated to Israel, including 37,105 tourists who changed their status once in Israel. This number is more than twice the total number of émigrés in the preceding 20 years. Altogether, some 1.2 million people had emigrated by the turn of the century, some 30% of the entire Jewish population of the former Soviet Union, and about twice the number of people who left Egypt in the Biblical Exodus.

The information on emigration from the Soviet Union comes from the files of the Hebrew Immigrant Aid Society in New York, the Union of Councils for Soviet Jews in Washington, D.C. (I am grateful to staffers at both these organizations who helped me dig out these data), and from *Shores of Refuge: a Hundred Years of Jewish Emigration* by Ronald Sanders (Henry Holt and Co., New York, 1988).

Appendix III

Text of SOS Petition

On 22 January 1980, our colleague Andrei Sakharov, an outstanding scientist and world-renowned leader of human rights, was arrested and exiled to Gorki by the Soviet authorities, for the "crime" of expressing his personal opinions. Since then he has been repeatedly harassed and even physically assaulted by the police. His wife reports he is in poor health. We must help!

To protest the Soviet government's savage treatment of their colleagues Orlov and Shcharansky, more than 2400 American scientists pledged last year to restrict their scientific cooperation with the Soviet Union. This action was strongly applauded by Sakharov and other Soviet dissidents (and was widely denounced by the Soviet media). Nearly 1000 French and Australian scientists have also adopted similar pledges. Because of Sakharov's exile and the deteriorating plight of other dissident scientists, we must act now and in much greater numbers than ever before.

We appeal to you, our fellow scientists and engineers the world over, to join together in a strong and significant protest of the Soviet Union's blatant violation of the human rights provisions of the Helsinki Accords to which it is a signatory. We propose a moratorium on scientific cooperation with the Soviet Union for a limited duration linked to Helsinki Accords actions.

To commemorate the founding of the Moscow Helsinki Watch Group by Orlov, Shcharansky and others, the Moratorium shall begin on the fourth anniversary of that date, 12 May 1980. Six months later, on 11 November 1980, there will commence a major conference in Madrid to monitor compliance with the Helsinki Accords, with representation from all 35 countries which signed the treaty. We propose to maintain the Moratorium until the end of the Madrid conference. Evidence from that meeting can then help determine the need for, and the course of, future action.

Scientists everywhere, acting independently of their governments, must express their deep concern now! We urge you to sign the pledge below and to solicit additional signatures from your professional colleagues. The pledge does not preclude personal communication with Soviet scientists in the interests of promoting human rights and world peace.

We will publicize the pledge, along with the names of signers, and send the list to Soviet President and Secretary Leonid Brezhnev and to the President of the Soviet Academy of Sciences, A. P. Aleksandrov.

Moratorium Pledge

To protest the human rights violations of the Soviet Union in the cases of Sakharov, Orlov, and Shcharansky, we, the undersigned scientists and engineers, pledge a moratorium on professional cooperation with the Soviet scientific community for a period beginning 12 May 1980, the anniversary of the founding of the Moscow Helsinki Watch Group, and ending at the completion of the November 1980 Madrid Conference to monitor the Helsinki Accords. During this period, we will not visit the Soviet Union or welcome Soviet scientists and engineers to our laboratories.

Appendix IV

Shcharansky's Closing Statement at His Trial

The following is a transcript of Anatoly B. Shcharansky's closing words at his trial on July 14, 1978, compiled from notes taken by his brother, Leonid.

In March and April, during interrogation, the chief investigators warned me that in the position I have taken during investigation, and held to here in court, I would be threatened with execution by firing squad, or at least 15 years. If I would agree to cooperate with the investigation for the purpose of destroying the Jewish emigration movement, they promised me early freedom and a quick reunion with my wife.

Five years ago, I submitted my application for exit to Israel. Now I am further than ever from my dream. It would seem to be cause for regret. But it is absolutely otherwise. I am happy. I am happy that I lived honestly, in peace with my conscience. I never compromised my soul, even under the threat of death.

I am happy that I helped people. I am proud that I knew and worked with such honest, brave, and courageous people as Sakharov, Orlov, Ginzburg, who are carrying on the traditions of the Russian intelligentsia. I am fortunate to have been witness to the process of the liberation of Jews of the U.S.S.R.

I hope that the absurd accusation against me and the entire Jewish emigration movement will not hinder the liberation of my people. My near ones and friends know how I wanted to exchange activity in the emigration movement for a life with my wife, Avital, in Israel.

For more than 2000 years the Jewish people, my people, have been dispersed. But wherever they are, wherever Jews are found, every year they have repeated, "Next year in Jerusalem." Now, when I am further than ever from my people, from Avital, facing many arduous years of imprisonment, I say, turning to my people, my Avital: Next year in Jerusalem.

Now I turn to you, the court, who were required to confirm a predetermined sentence. To you I have nothing to say.

CHRONOS
BOOKS

HISTORY

Chronos Books is an historical non-fiction imprint. Chronos publishes real history for real people; bringing to life people, places and events in an imaginative, easy-to-digest and accessible way - histories that pass on their stories to a generation of new readers.
If you have enjoyed this book, why not tell other readers by posting a review on your preferred book site.

Recent bestsellers from Chronos Books are:

Lady Katherine Knollys
The Unacknowledged Daughter of King Henry VIII
Sarah-Beth Watkins
A comprehensive account of Katherine Knollys' questionable paternity, her previously unexplored life in the Tudor court and her intriguing relationship with Elizabeth I.
Paperback: 978-1-78279-585-8 ebook: 978-1-78279-584-1

Cromwell was Framed

Ireland 1649

Tom Reilly

Revealed: The definitive research that proves the Irish nation owes Oliver Cromwell a huge posthumous apology for wrongly convicting him of civilian atrocities in 1649.

Paperback: 978-1-78279-516-2 ebook: 978-1-78279-515-5

Why The CIA Killed JFK and Malcolm X

The Secret Drug Trade in Laos

John Koerner

A new groundbreaking work presenting evidence that the CIA silenced JFK to protect its secret drug trade in Laos.

Paperback: 978-1-78279-701-2 ebook: 978-1-78279-700-5

The Disappearing Ninth Legion

A Popular History

Mark Olly

The Disappearing Ninth Legion examines hard evidence for the foundation, development, mysterious disappearance, or possible continuation of Rome's lost Legion.

Paperback: 978-1-84694-559-5 ebook: 978-1-84694-931-9

Beaten But Not Defeated

Siegfried Moos - A German anti-Nazi who settled in Britain

Merilyn Moos

Siegi Moos, an anti-Nazi and active member of the German Communist Party, escaped Germany in 1933 and, exiled in Britain, sought another route to the transformation of capitalism.

Paperback: 978-1-78279-677-0 ebook: 978-1-78279-676-3

A Schoolboy's Wartime Letters
An evacuee's life in WWII — A Personal Memoir
Geoffrey Iley
A boy writes home during WWII, revealing his own fascinating story, full of zest for life, information and humour.
Paperback: 978-1-78279-504-9 ebook: 978-1-78279-503-2

The Life & Times of the Real Robyn Hoode
Mark Olly
A journey of discovery. The chronicles of the genuine historical character, Robyn Hoode, and how he became one of England's greatest legends.
Paperback: 978-1-78535-059-7 ebook: 978-1-78535-060-3

Readers of ebooks can buy or view any of these bestsellers by clicking on the live link in the title. Most titles are published in paperback and as an ebook. Paperbacks are available in traditional bookshops. Both print and ebook formats are available online.

Find more titles and sign up to our readers' newsletter at http://www.johnhuntpublishing.com/history-home

Follow us on Facebook at https://www.facebook.com/ChronosBooks

and Twitter at https://twitter.com/ChronosBooks